A Sports-Oriented Approach to Introductory Statistics

First Edition

By Andrew Wiesner
Pennsylvania State University

Bassim Hamadeh, CEO and Publisher
Christopher Foster, General Vice President
Michael Simpson, Vice President of Acquisitions
Jessica Knott, Managing Editor
Kevin Fahey, Cognella Marketing Manager
Jess Busch, Senior Graphic Designer
Melissa Barcomb, Acquisitions Editor
Sarah Wheeler, Senior Project Editor
Stephanie Sandler, Licensing Associate

First published in the United States of America in 2013 by Cognella, Inc.

Printed in the United States of America

ISBN: 978-1-60927-289-0 (pbk)/ 978-1-60927-289-0 (br)

www.cognella.com 800.200.3908

Contents

Introduction

Frequently students enrolled in introductory statistics courses are apprehensive about taking the class. They fear the worst and hope for the best. As the title suggests, this book brings a unique approach to the teaching and learning of introductory statistics, where the focus is mainly on helping the student. By using sports as a theme, students can better understand and appreciate the general topics covered in such a course. From data description through simple linear regression, each topic is discussed and explained via examples from the sports world. However, the reader need not be well-versed in sports in order to gain value from this book. They simply need to have a general interest in sports, primarily major league baseball, football, basketball or hockey. Data from these four leagues serve as the primary source in the discussions. Yet there is a sprinkling of examples from college football and basketball, plus professional golf. When specific individuals are referenced, the names are often popular ones, such as LeBron James or Alex Rodriguez.

The book can also assist students at various stages in their education. For instance, the high school student taking AP statistics will find this text to be a useful supplement. So, too, will the undergraduate student taking their introductory or elementary statistics course; or the graduate student looking to refresh basic statistical concepts prior to taking on a more challenging course in statistics. Although this text is not attached to a specific course-required textbook, the material covered here will align itself with the topics covered in almost every introductory statistics course.

Other than using a sports-theme, what makes this book different from the standard textbook? Mainly, this book does not contain end of chapter problems. Instead, throughout each chapter are examples accompanied by complete, step-by-step solutions. That means no problems where the answer is a simple one word or letter provided in some index. Nor do you have to purchase a separate solution manual that contains answers to all of the problems. Symbols and formulas used within a chapter are organized separately at the end each chapter for easy reference. Overall, this arrangement allows for fewer pages resulting in a lower cost. Additionally, this makes for easier reading. The examples, when practical, are supported with annotated output or graphics using the statistical software Minitab.

For teachers, by suggesting the book as a supplemental item you provide your students with an additional, low cost opportunity to become engaged in the course while increasing their understanding. You will find the topics inside will follow along with those in your teachings. You may already be familiar with your students having a keen interest in sports. For instance, during meetings with a student(s) to discuss the test of one mean, the student may cite some sport as a topic of interest. In response, you would attempt to construct a one-mean example from this sport. That scenario reflects the impetus behind this book. You now have a series of interesting examples—to many of your students—from which to draw.

In the end, we hope you find this textbook enjoyable, easy to read, and educational. For students, an improvement in your grade and overall understanding of basic statistics would be welcomed, too! As with most statistics texts, the first few chapters are more theory- or math-based: Consider these chapters the "offensive linemen" in statistics. Just like the linemen are often overlooked or underappreciated, these chapters frequently present the biggest hurdle for students. Yet they provide the necessary support in understanding what comes next. The later chapters covering applications of statistical methods reflect the "skill players." This material is more exciting, as this is where we learn, for example, how to predict points in the NHL, determine if there is a home field advantage in football, or compare MLB teams to find out if statistically any one team has been the worst over the past ten years. So give the book a read; maybe you will learn some techniques that enhance your opinion when "discussing" some sports topic with friends.

Acknowledgments

I am indebted to several people for whom without their input this endeavor would never have taken shape. First, to Brett Lissenden a former student and a current PhD candidate in Economics at the University of Virginia. Brett offered invaluable insight and suggestions while reviewing each chapter during the writing process. Second, to my brother Karl and sisters Ruth and Eileen, who served as the most influential people in developing my interest in sports and sustaining me in my youth. They made me earn my victories. Third, to my parents Fran and Rita. Beyond the obvious reason—without them I would not exist—they were always there to support my athletic whims, regardless if I won (some) or lost (more). Finally, to my wife Kim of whom I could say much but will simply say this: I could not have found a better person with whom to share my life. Each day with her is a reminder that I have won the lottery.

Describing Data with Numbers and Graphs

1

1.1: Types of Data

This is possibly the most fundamental aspect of undergraduate statistics. Many topics begin by first identifying the variable which is being studied. In Chapter 1, the choice of graphs and numerical summaries will be dependent on variable type. In subsequent chapters, the variable in question will provide the starting block to selecting the proper statistical method in which to further analyze the data. Understanding variable types is extremely important and valuable.

Categorical (or Qualitative) and Quantitative (or Measurement)

The easiest way to decide if a variable should be classified as categorical or quantitative is to simply answer this: "How would I answer if asked?" If your answer would be a "text" or word, then the variable is categorical; if your answer would be a number, then it would be quantitative. Here are some examples; think how you would answer:

1. What is your favorite sport?
2. What is your favorite NFL team?
3. How many touchdown passes did Tom Brady throw in 2011?
4. Did your team win the 2012 Super Bowl?
5. Who was the heaviest NFL player in 2010?
6. How much did the heaviest 2010 NFL player weigh?

Now consider your responses (below are mine), and classify the variable based on your response.

1. Golf (to play), football (to watch).
2. Steelers.
3. 39.
4. No.

5. Aaron Gibson of the Bears.

6. Aaron Gibson weighed **375** pounds!

Any of the answers requiring a text response would be labeled *categorical* (i.e., 1, 2, 4, and 5), while those prompting a numerical response are *quantitative* (i.e., 3 and 6). Notice also the difference between examples 5 and 6. In 5, our interest was in *who*, which implies a name, but in 6 our interest was *how much*, which elicits a numeric response.

Common Mistakes

This all seems fairly simple, so where might you become confused? There are two primary reasons a student confuses variable types, and typically the error comes from labeling a categorical variable as quantitative. This occurs when:

A. The data is summarized and a number is associated with the category; and

B. The categories are represented by a number

Let us consider the first point. You are at a Super Bowl party for the Steelers-Packers game and the question goes out, "Who is everyone rooting for?" The responses come in as either Steelers or Packers, and the results are expressed as follows: 18 people are rooting for Steelers, 12 for the Packers. And now the confusion begins. You see the numbers 18 and 12 and automatically think "quantitative," forgetting the original question of *who* are you rooting for. The numbers are just summaries of the categorical responses. That is, when you were asked the question, would you have responded with "eighteen?" Of course not; you would have offered a team name.

Point two can be less obvious but just as troublesome. As you start working with data—either on your calculator or some statistical software—you will often be required to use numbers instead of words. That is, you will have to use a number to represent a category. For example, you want to research whether past World Series champions swept the series or not. For starters, recognize this as a categorical reference, as the answer is either Yes or No. However, try putting Yes/No in your calculator and having this counted! To use these responses, you may need to relabel them, say, 0 for no and 1 for yes. Then you can count the number of 1"s to come up with how many teams swept a World Series (the answer is 20!). Where students make the mistake is "seeing" the 0 and 1 and thinking, "Ah, numbers—must be a quantitative variable." They forget that these numbers just represent a category and are arbitrary; that is, we could have assigned 10 as No and 100 as Yes and we still would have counted 100 twenty times.

Classifications of Categorical and Quantitative Variables

Once we have defined the variable as either categorical or quantitative, the next step is to determine the *classification* of the categorical or quantitative variable. For categorical variables, we have three basic classes:

1. Binary or Dichotomous (only two category levels).
2. Nominal (more than two levels but without one level being "superior" to another).
3. Ordinal (the levels have some hierarchy or "superiority"; one can be ranked above another).

Basic examples of binary categorical variables are MLB or NFL Conference (American, National). In each situation there are only two possible choices.

Using the above MLB and NFL reference, the divisions within each of these sports would be classified as nominal. Each division is considered equal to all the others; there is no hierarchy or superiority between the divisions. Even though you may favor one over the other doesn't make that one superior to the other! That is, just because you love the Phillies and hate the Pirates doesn't make the National League East "better" than the National League Central.

As to ordinal, consider the "highest level one played" for a particular sport. For instance, if you were taking a survey where the question posed was, "What was the highest level for which you participated in football?" the possible choices could be: None, Pee-Wee, High School, College, or Professional. From these categories you can clearly see a "hierarchy" or "superiority" in the levels; someone who played professionally would certainly be considered having achieved a higher level of competition than someone who only reached the college level, etc. Consider even your high school days, where your state may have had classification for sports (e.g., in Pennsylvania there are four classifications: A, AA, AAA, and AAAA, with A being the smallest and AAAA being the largest class). These, too, would be thought of as ordinal. Note, this shouldn't be interpreted as necessarily one level always being better than another (i.e., an A-level team could never beat an AAAA-level team), but that these levels are based on some form of achievement (in the case of level of play) or size (number of students enrolled determining school classification).

1.2: Numerical Representation - Center and Spread

Data can best be described in two ways: the **central tendency** (or center) and **variability** (or spread).

Center

The terms **mean**, **median**, and **mode** are already commonly used words. Yet when someone asks you to give the *average*, the initial (and popular) response is to offer the mean. However, in statistics the word average can often relate to any one of the three common descriptors: **mean**, **median**, or **mode.** The first distinction is to understand for which situations should one pick mean, median, or mode? To start, the mean and median are most appropriate to use when the data is *quantitative*. This leaves mode to be best applied when the data is *categorical*.

The **mode** is the easiest of the three to calculate and to explain: It is simply the category level with the most observations or highest frequency. If there is only one mode (i.e., one category level with the most observations), this is referred to as **unimodal**. If there is more than one mode (i.e., more than one category level with the most observations), this is called **multimodal**. Within multimodal, there are two common expressions: **bimodal** when there are two modes and **trimodal** when there are three modes. Considering the World Series winners previously mentioned, there would be one mode: the Yankees with their 27 championships. We will see below examples of multimodal data sets.

With the **mode** representing the "most frequent observation," this is found by calculating the number of times each unique data point appears. For example, the 2011–2012 NCAA season, these heights (in inches) were as follows (courtesy www.goduke.com):

Men: 76, 72, 73, 82, 80, 79, 79, 76, 82, 74, 83, 83, and 81.

Women: 75, 69, 67, 71, 70, 72, 74, 76, 73, 75, 66, and 77.

For the Duke men, the heights of 76, 79, 82, and 83 inches each appear twice making this set of data multimodal. As for the Duke women, the height of 75 inches appears most frequently—twice—making this set of data unimodal.

The **mean** of a set of data is referred to as the mathematical average found by adding up all the data points and dividing by the total number of data points we added. A very straightforward process and one most people understand early in life! Where confusion sometimes comes into play is when a formula is involved. Some common expressions used to represent the mean are:

$$\frac{\sum_{i=1}^{n} X_i}{n} \text{ or } \frac{\sum_{i=1}^{n} x_i}{n} \text{ or } \frac{\sum_{i=1}^{n} Y_i}{n} \text{ or } \frac{\sum_{i=1}^{n} y_i}{n}$$

Whether you use capital letters (X, Y) or lower-case letters (x, y) doesn't really matter. Just consider the lettering as shorthand. This "*n*" represents the **sample size** or the number of observations for our data. The X (or Y) is shorthand for the variable in question. For instance, we could say, "Let X represent the heights of the Duke men's basketball team," or "Let Y represent the heights of the Duke women's basketball team." This shorthand helps when we see the formula because it directs us to the proper variable.

Returning to the Duke basketball heights, this is where the shorthand is helpful. Instead of asking, "What is the mean heights of the Duke men's basketball team," one could simply say "What is the mean of X?" The two questions are identical and the labeling is important to keep us from finding the wrong mean from what was asked (e.g., finding the women's mean instead of the men's mean). To answer this question for the mean of the men's team, the sample size (n) is 13 and "$i = 1$" would have us start at the first listed height of 76 inches. We would proceed to sum all of the male heights, 1020 inches. Since we added 13 heights, we would divide the sum by 13 to get the mean height of the men's team: 78.5 inches.

The mean for the women's team would be found in similar fashion (this would be finding the mean for Y), except notice that there are only 12 women's heights making the sample size, n, 12. Starting with the first height of 75, we get a sum of 865 inches. We divide this sum by 12 to get the mean height of the women's team of 72.1 inches.

The last bit of information is the symbol used to represent the mean for a set of data. This is most often represented by putting a bar over the letter representing the variable of interest. That is, the symbol for the men's mean would be $\overline{X}$ (read "X bar") and for the women's mean this would be $\overline{Y}$ (read "Y bar"). Putting this all together, for the mean height of the men's basketball team at Duke, we would have $\overline{X} = 78.5$ inches; for the mean height of the women's basketball team at Duke, we would have $\overline{Y} = 72.1$ inches.

Special Note: The entire set of observations makes up the sample; the values within the data make up the individual observations of the sample. For example, the minimum male height of 72 inches is not the sample but is a single observation within the sample.

Median

The **median** of a set of data is referred as the midpoint. As with the mean, the value of the median does not necessarily have to be a value observed within the set of observations. The principle behind the median is to "split the data in half," with half of the observations falling at or below this point (in turn, about half are at or above this point). The *key* to finding the median is that the data must be ordered from the minimum to maximum values. The steps to finding the median are as follows:

1. Order the data from minimum to maximum.
2. Find the *location point* of the median by taking the number of observations plus one and dividing by two. This is often represented by $\frac{n+1}{2}$.
3. Calculate the value within the ordered data from Step 1 that represents the location point from Step 2. This value is the **median**.

We can continue to use the Duke basketball teams as an example.

Step 1: Order the data from minimum to maximum.

Men: 72, 73, 74, 76, 76, 79, 79, 80, 81, 82, 82, 83, 83.

Women: 66, 67, 69, 70, 71, 72, 73, 74, 75, 75, 76, 77.

Step 2: Find the location point using $\frac{n+1}{2}$.

Men: $\frac{13+1}{2} = \frac{14}{2} = 7$

Women: $\frac{12+1}{2} = \frac{13}{2} = 6.5$

Step 3: Use location point and find median within ordered data.

Men: 72, 73, 74, 76, 76, 79, **79**, 80, 81, 82, 82, 83, 83.

Women: 66, 67, 69, 70, 71, 72,| 73, 74, 75, 75, 76, 77

Men: Location point "7" means the 7th observation is the median. Starting at 72 the seventh observation is 79 inches.

Women: Location point "6.5" means the median is between the 6th and 7th observation, illustrated above by |. Starting at 66, the sixth observation is 72 and the seventh is 73. The median is the average or mean of these two observations, resulting in a median of 72.5 inches.

TIP: The reason two examples were provided with one sample size being even and the other odd was to point out a tip in calculating the position of the median. When the sample size is *even*, the location point will result in averaging two values. That is, the location point will always end with the decimal ".5"—although the median itself won't necessarily end in ".5." The median, therefore, may or may not be an observed value in the data. However, if the sample size is *odd*, then the location point will be a whole number and the median will be a value observed in the data.

Interpreting Mean and Median

As you can see the calculations for the mean and median are not difficult. But what about the interpretations? The **mean** can be interpreted as the "mathematical average," or **balancing point**, while the median is considered the fifty-fifty point. In the example above, the median male height of 79 inches would be interpreted as, "Fifty percent of the players of the Duke men's basketball team in 2011–2012 had a height of 79 inches or less," while the interpretation for the median female heights would be, "Fifty percent of the players of the Duke women's basketball team in 2011–2012 had a height of 72.5 inches or less."

Common Confusion Interpreting Median

Among students there is sometimes confusion when interpreting the median (and another statistic called the quartiles, which is coming up). As you can see with the previous example, the median is not always a value that is observed in the data: The 72.5 is not observed in the women's heights. However, this doesn't change the interpretation. If you examine the ordered list of heights for the men, you will find that seven heights are at or below the median 79 and seven heights are at or above this median. This is true for the women's heights, as six heights are at and below the median of 72.5 and six heights are also at or above this height. Don't be confused with the "at or above," especially in regard to the second example. Think more of "at or before." Yes, there are no heights at 72.5 inches but that is not the point. For instance, what if all 12 players on the women's team were 72 inches tall? The ordered string with median location marked would be as follows:

72, 72, 72, 72, 72, 72,| 72, 72, 72, 72, 72, 72

The median would be 72 inches, implying half (i.e., six) observations come *before* 72 and six come *after*. As you can see, all the observations are 72: none are before and none are after. However, if you include the phrasing "at or below," then this would be more reflective of this particular situation.

Better Choice: Mean or Median?

When we have quantitative data, we have a choice between using the mean or the median to represent the average. The question becomes: Which one is better? Well, if one of them was always better, there wouldn't be the need for the second. Therefore, there must be circumstances when the mean is preferred and vice versa. The decision is dependent on the shape of the data. When the data is bell shaped, the mean is a better choice, since the mean incorporates all of the values in the calculation. However, if the shape is skewed, the median offers a better measurement of the average. Without taking into account all observations, the median is less affected by outliers, or extreme observations, that cause the data to be skewed. This makes the median a more **robust** or **resistant** estimator of the average. In an upcoming section of Chapter 1 we will further discuss defining data shapes.

Spread

Several measures exist to examine the spread of a data set, and similar to the mean and median, this should be reserved for use with **quantitative data**. The most common are:

1. Range.
2. Variance and standard deviation (these are intertwined and studied together).
3. Interquartile range (IQR), percentiles, and quartiles.

Range

The **range** of a data set is simply the difference between the maximum and minimum value. Simple enough, but confusing to some nonetheless. Why? Because if you were asked to find the range for the Duke men's basketball team, would you answer 72 to 83 inches, or answer 11 inches? A common response is to say 72 to 83, but in statistics we use the *difference* to represent the range. This makes the answer, "The range is 11 inches." Well, that seems strange, doesn't it, so why use that approach? One reason for using the difference to define the range is to make comparisons across data sets easier. For example, which would be easier (think quicker!) to use in order to compare the ranges of the Duke men's and women's basketball teams? To provide the minimum and maximum values or the range as defined here? Take a look, and you decide which is simpler to compare.

Min to max: Men is 72 to 83 and Women is 66 to 77.

Range: Men and Women are both 11.

Variance and Standard Deviation

The variance and standard deviation are used in conjunction with each other, simply because once the **variance** is calculated, the **standard deviation** can be found by taking the square root of the variance. This makes for the following relationship:

Standard Deviation = $\sqrt{Variance}$ or using more common notation: $S = \sqrt{S^2}$ where **S** represents the standard deviation and $\mathbf{S^2}$ is the variance.

The more popular of the two is the standard deviation. In fact, the standard deviation is the most popular measure of data spread. The reason is that the units used in the data (e.g., pounds, inches, dollars) will be the same units for the standard deviation. Also, the standard deviation uses all of the data in its calculation.

Formula:

$$\text{Variance or } S^2 = \frac{\sum_{i=1}^{n}(X_i - \bar{X})^2}{n-1}$$

$$\text{Standard deviation or } S = \sqrt{\frac{\sum_{i=1}^{n}(X_i - \bar{X})^2}{n-1}}$$

The expression $(X_i - \bar{X})$ is called a **deviation** and represents the difference between each observation, X_i, and the mean of the observations, Because of this, some deviations will be positive (data points greater than the mean) and others negative (data points below the mean). The concept of deviations helps to explain the reasoning behind the mean being considered the balancing point: The negative and positive deviations *counterbalance* each other. As a result, *the sum of the deviations will equal zero*!

REMINDER: As with the formula for calculating the mean, don't get confused by the lettering! The remaining notations are similar to those found for calculating the mean.

Interpretation of Standard Deviation

A common interpretation is that the standard deviation is the average or typical distance the observations in the data fall from the mean of the data. This implies that the larger the value of the standard deviation S, the wider the spread of the data.

Application of Formula

Using the data for the starting offensive lineman for the Penn State football team in 2011 (courtesy www.gopsusports.com) the weights, in pounds, are:

299, 304, 293, 310, 294.

First, we calculate the mean by summing up the data (1500) and then divide by five, the number of data points summed. This results in a mean of 300.

Second. we calculate the variance S^2.

$$S^2 = \frac{(299-300)^2+(304-300)^2+(293-300)^2+(310-300)^2+(294-300)^2}{5-1}$$

$$S^2 = \frac{(-1)^2+(4)^2+(-7)^2+(10)^2+(-6)^2}{4} = \frac{1+16+49+100+36}{4} = \frac{202}{4} = 50.5$$

So the variance S^2 would be 50.5 pounds^2.

Third, to get the standard deviation, S we take the square root of the variance S^2.

$S = \sqrt{50.5} = 7.1$ of the standard deviation is 7.1 pounds.

Special Notes:

1. Remember from basic algebra that the square root of a number can either be positive or negative. Even though the standard deviation is typically written using only the positive value, keep in mind that the meaning includes "plus and minus." This means that sometimes the standard deviation may be seen as $\pm$ S. For example, we could write the standard deviation for this example as S = $\pm$ 7.1, indicating that the typical observation, on average, is roughly 7.1 greater or 7.1 less than the mean.
2. If we sum the deviations -1, 4, -7, 10, and -6, we get zero, a property of deviations.

Special Application of Standard Deviation: The Empirical Rule or 68-95-99.7 Rule

If the **shape**, also called **distribution**, of the data is bell shaped or nearly bell shaped, then a unique concept exists that allows us to use the standard deviation to approximate the percentage of the data that falls within a certain number of standard deviations from the mean. This unique concept is referred to as the **Empirical Rule**, or the **68-95-99.7 Rule**.

In general, if we know that the shape of the data is at least close to bell shaped, then the **Empirical Rule** states the following: Approximately

68% of the observations will fall within **1 standard deviation** of the mean:

95% of the observations will fall within **2 standard deviations** of the mean:

99.7%, about all, of the observations will fall within **3 standard deviations** of the mean:

Example: Consider the 2009 average points per game for the 32 NFL teams. The data, courtesy of www.espn.com, is:

31.9	29.4	28.8	28.4	26.8	26.7	26.0	25.1
24.4	24.2	23.4	23.0	22.7	22.6	22.5	22.1
21.8	20.6	20.4	20.4	19.7	19.1	18.4	18.1
17.5	16.6	16.4	16.1	15.3	15.2	12.3	10.9

The mean points per game is 21.5 with 5 being the standard deviation.

Figure 1.1 presents a histogram of the data fitted with a distribution or shape which is approximately bell shaped, with a mean of 21.5 and standard deviation of 5.

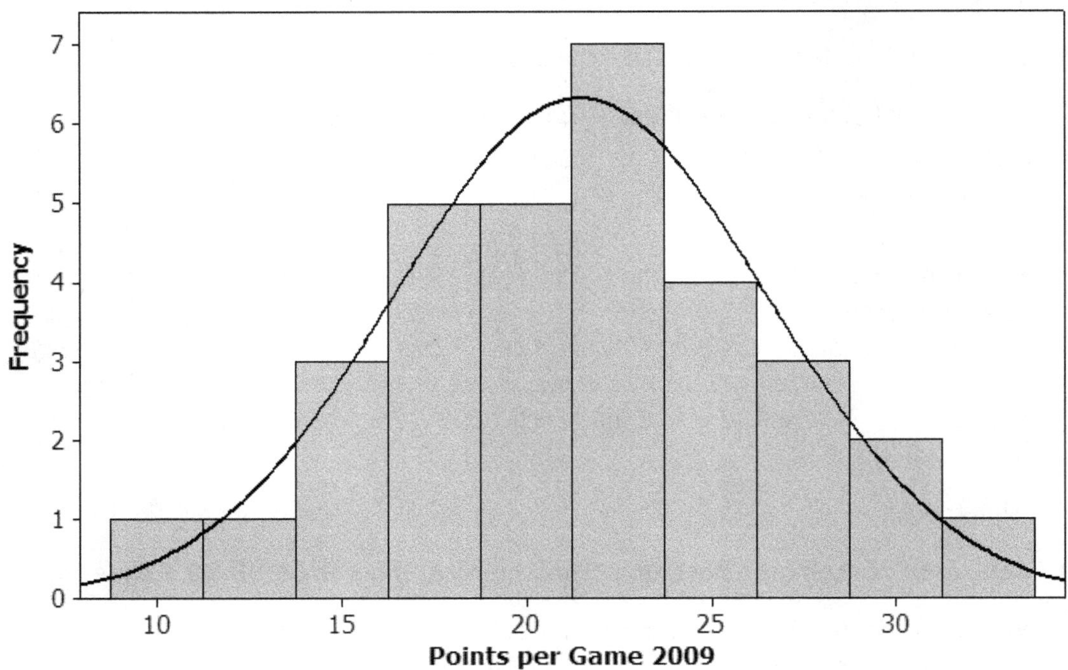

With there being 32 observations, according to the Empirical Rule, approximately:

68% of the 32 teams should average between 21.5 ± 5 or 16.5 to 26.5 points per game.

95% of the 32 teams should average between 21.5 ± 10 or 11.5 to 31.5 points per game.

99.7% of the 32 teams should average between 21.5 ± 15 or 6.5 to 36.5 points per game.

How did the rule perform? Table 1.1 compares the results to the Empirical Rule:

Empirical Rule	Range	Teams in Range	Percent
68%	16.5 to 26.5	20	63%
95%	11.5 to 31.5	30	94%
99.7%	6.5 to 36.5	30	100%

TABLE 1.1: Empirical Rule application to 2009 NFL points per game data.

Not bad! Note that with a larger sample, we would expect the breakdown of percentages to be even closer to those defined by the Empirical Rule.

Common Mistake: A common error made by students when using the Empirical Rule is forgetting to include the standard deviation. That is, they simply add and subtract the numbers 1, 2, and 3, respectively from the mean instead of using S, two times S, and three times S. Remember, the rule applies to a set number of standard deviations from the mean. For a simple pictorial representation of how the Empirical Rule works see Figure 1.2.

FIGURE 1.2: Empirical Rule.

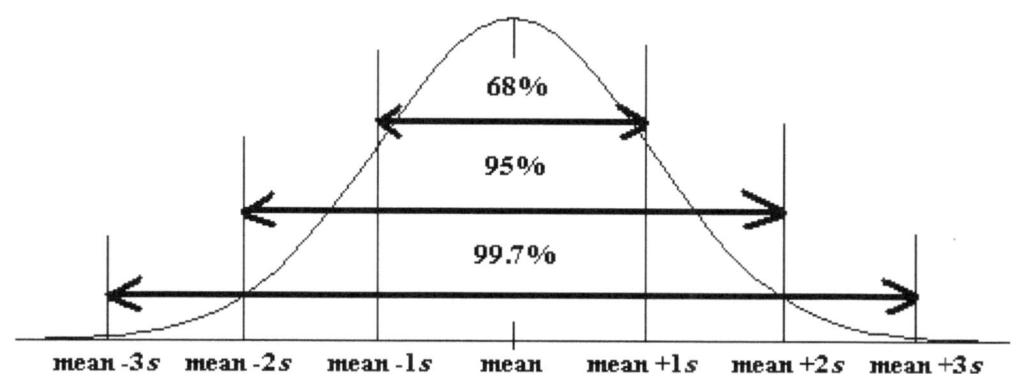

Interquartile Range (IQR), Percentiles, and Quartiles

The last measures of spread are the interquartile range (IQR), percentiles, and quartiles, starting with percentiles.

Percentiles are easy to interpret as they simply represent *the percent of the observations that fall at or below that value*. We have already discussed on specific percentile: the median. The median, if you recall, splits the data in half, making the median the 50th percentile: For a set of data, 50 percent of the observations fall at or below the median. There are three helpful percentiles called **quartiles**.

When thinking of **quartiles**, a word that should come to mind is "quarters"—and for good reason. **Quartiles** split the data into four roughly equal sections of 25%. That means the data is split into four quartiles, each one roughly covering 25 percent more observations than the previous quartile. Thus we have Quartile One (Q1), Quartile Two (Q2), Quartile Three (Q3), and Quartile Four (Q4), with interpretations as follows:

Q1: Approximately 25% of the observations fall at or below this value, and approximately 75% fall at or above this value.

Q2: Approximately 50% of the observations fall at or below this value, and approximately 50% fall at or above this value. Q2 has a special name: **median**.

Q3: Approximately 75% of the observations fall at or below this value, and approximately 25% fall at or above this value.

There is usually no mention of a fourth quartile as this would be the maximum value with all data points falling at or below the maximum. So instead of Q4 we have the maximum.

Finding Quartiles

Surprisingly, there is more than one way to calculate quartiles: One method involves finding the median of the lower and upper halves; the second method—typically reserved for software—uses interpolation methods. For our purposes, we will examine the first method, as both methods arrive at roughly the same values.

1. Order the data from minimum to maximum.
2. Find the location points for Q1 using $\frac{n+1}{4}$ and Q3 using $3 * \left(\frac{n+1}{4}\right)$ **NOTE**: If these values come to a decimal point other than .5, just use .5 for easier calculations.
3. Calculate the value within the ordered data from Step 1 that represents the location points from Step 2. These values are **Q1** and **Q3**, respectively.

Explanation: Can you see the logic behind these formulas for the location points? Recall for the median we used $\frac{n+1}{2}$ and the median gives us the halfway point. The first and third quartiles provide the 1/4 and 3/4 points, respectively. Putting this together, you notice that for each of these quartiles, we simply multiplied n + 1 by the quartile. That is, the location points for Q1 is found by 1/4 * (n + 1); for Q2 the median 1/2 * (n + 1); and for Q3 we use 3/4 * (n + 1).

Example: We will demonstrate finding the first and third quartiles using the Duke heights data already arranged in order.

Men: 72, 73, 74, 76, 76, 79, 79, 80, 81, 82, 82, 83, 83.

Women: 66, 67, 69, 70, 71, 72, 73, 74, 75, 75, 76, 77.

Step 2: Find the location points.

	Q1 location point	Q3 location point
Men:	$\frac{13+1}{4} = \frac{14}{4} = 3.5$	$3 * \left(\frac{13+1}{4}\right) = 3 * \frac{14}{4} = 3 * 3.5 = 10.5$
Women:	$\frac{12+1}{4} = \frac{13}{4} = 3.25$ (*use* 3,5)	$3 * \left(\frac{12+1}{4}\right) = 3 * \frac{13}{4} = 3 * 3.25 = 9.75$

Step 3: Find location points to get quartiles. These are marked below with |.

Men: 72, 73, 74, | 76, 76, 79, 79, 80, 81, 82, | 82, 83, 83.

Women: 66, 67, 69, | 70, 71, 72, 73, 74, 75, | 75, 76, 77.

Men: Q1 location of 3.5 has the first quartile being the average of third and fourth observation, or Q1 = 75. Q3 location of 10.5 has the first quartile being the average of tenth and eleventh observation, or Q3 = 82.

Women: Q1 location of 3.5 has the first quartile being the average of third and fourth observation, or Q1 = 69.5. Q3 location of 9.5 has the first quartile being the average of ninth and tenth observation, or Q3 = 75.

Interpretation Q1: The first quartile, Q1, would have the following interpretation:

Men: About 25% of the players have heights at or below 75 inches, and about 75% of the players have heights at or above 75 inches. Put another way, about 25% of the players are at most 75 inches tall, and about 75% of the players are at least 75 inches tall.

Women: About 25% of the players have heights at or below 69.5 inches, and about 75% of the players have heights at or above 69.5 inches. Put another way, about 25% of the players are at most 69.5 inches tall, and about 75% of the players are at least 69.5 inches tall.

Interpretation Q3: The third quartile, Q3, would have the following interpretation:

Men: About 75% of the players have heights at or below 82 inches, and about 25% of the players have heights at or above 82 inches. Put another way, about 75% of the players are at most 82 inches tall, and about 25% of the players are at least 82 inches tall.

Women: About 75% of the players have heights at or below 75 inches, and about 25% of the players have heights at or above 75 inches. Put another way, about 75% of the players are at most 75 inches tall, and about 25% of the players are at least 75 inches tall.

Common Confusion Interpreting Quartiles and Percentiles

Going through the examples, the process can seem very straightforward. However, there are interpretive errors that often arise. The confusion sets in when students are asked to explain the quartiles when the language includes phrasing such as *at most* and *at least*. To keep this straight, bear in mind that "at most" implies a max point on where to stop, and "at least" implies a minimum point on where to begin. An easy way to remember this is to think of any sport you may have played, including intramurals, where there was a minimum and maximum number of players needed in order to play the game. For example, in many IM football games, you can play with a maximum of 11 players (i.e., at most, 11 on the field at one time), but you need a minimum of 8 players (that is, at least 8 on the field at one time). Here is a quick reference guide:

Q1: 25% at most or 75% at least **Q3:** 75% at most or 25% at least

Interquartile Range or IQR

The **interquartile range**, also called the **IQR**, is another measure of spread and represents the range of values for the **middle 50 percent**. The IQR is the difference between the first and third quartiles, or

IQR = Q3 - Q1

Example: From the Duke heights data and the quartiles found above, the IQR values would be:

Men: IQR = 82 - 75 = 7 Women: IQR = 75 - 69.5 = 5.5

Interpretation: For the men, this would be interpreted as "approximately half of the players had heights from 75 to 82 inches," and for the women, "approximately half of the players had heights from 69.5 to 75 inches."

The IQR provides the middle fifty percent, since on either we are examining the difference between Quartiles One and Three. This results in a quarter (25%) of the observations falling onto either side of the IQR.

SPECIAL NOTE: The IQR and median are NOT the same! The median is a *quartile*, and gives the point from the minimum value to the median, which is the *first* fifty percent. The IQR gives the *middle* fifty percent, as it is the *difference* between two quartiles.

1.3 Finding Outliers Using IQR

What are outliers? If you had to guess, you would probably come up with a definition along the lines of an outlier being an observation that doesn't appear to belong with the rest of the data. This would be accurate, as an **outlier** is an unusual observation, or one that appears to be somewhat distant from the rest of the data. And there can be more than one outlier in a data set! Imagine a conversation with a friend where you are debating a player's performance. Each of you uses data to support your cause, and the rebuttal is, "Well that's an outlier!" For instance, in the fall of 2011, Georgia Southern scored 21 points against Alabama in football. The Tide was known to have a great defense, but one may speculate on how good it could be to give up 21 points to Georgia Southern. A Tide supporter might reply, "That 21 points was an outlier; look how we did against everybody else! Our defense was tired that day or looking ahead to the next game against Auburn." So how does someone go about classifying an observation as an outlier? One method is to use the IQR.

To find outliers, we first use the IQR to set up a "fence" around our data. After building this fence, we compare our data to this fence. Any observations outside this fence are considered outliers.

Building the Fence: The fence consists of a lower and upper bound or **fence point**. These lower and upper points are found as follows:

Lower: Q1 - 1.5* IQR **Upper:** Q3 + 1.5*IQR

Example: Applying the method to the Duke heights data:

Men: 72, 73, 74, 76, 76, 79, 79, 80, 81, 82, 82, 83, 83.

Women: 66, 67, 69, 70, 71, 72, 73, 74, 75, 75, 76, 77.

Men:

Lower: 75 - 1.5* 7 = 75 - 10.5 = 64.5 Upper: 82 + 1.5*7 = 82 + 10.5 = 92.5

With all height observations for the men falling within this fence, no heights would be considered unusual or an outlier.

Women:

Lower: 69.5 - 1.5* 5.5 = 69.5 - 8.25 = 61.25 Upper: 75 + 1.5*5.5 = 75 + 8.25 = 83.25

Again, with all height observations for the women falling within this fence, no heights would be considered unusual or an outlier.

Question: Were the 21 points Alabama gave up to Georgia Southern an outlier?

Answer: The points surrendered by Alabama in their 12 regular season games were (ordered):

0, 0, 6, 7, 7, 7, 9, 10, 11, 14, 14, 21.

The first and third quartile positions, respectively, are and resulting in Q1 of 6.5 - average of 6 and 7 points, and Q3 of 12.5 - average of 11 and 14 points.

IQR = Q3 - Q1 = 12.5 - 6.5 = 6

Lower Fence = Q1 - 1.5*IQR = 6.5 - 1.5*6 = 6.5 - 9 = - 2.5 (obviously a team cannot score "negative" points, meaning 0 would be the realistic lower fence)

Upper Fence: Q3 + 1.5*IQR = 12.5 + 1.5*6 = 12.5 + 9 = 21.5

Since all observations—the points Bama allowed—fall on or within these two fence points, there are no outliers. So the answer to the question is, no; the 21 points given up to Georgia Southern, though the most allowed, was not an outlier using the IQR method.

Five Number Summary

The **five number summary** for a set of data consists of exactly that: five numbers. The five that make up the summary are the minimum, Q1, median, Q3, and maximum values. Notice that this summary does NOT include the IQR, the sample size, standard deviation, mean, or range. From our Duke data, the five number summaries are:

Men: 72, 75, 79, 82, and 83 Women: 66, 69.5, 72.5, 75, and 77

Common Mistake: The five number summary seems so easy—what mistake could one possibly make? Simple, you overthink! The five number summary is exactly what it says it is: five numbers. So your answer can only consist of five numbers! Another mistake is that you include the wrong five. Remember that the five number summary is precisely that: a summary of the observation values. The five we use are: minimum, Q1, median, Q3, and maximum.

1.4 Graphing Data

Many graphs exist to represent data, but we are only going to consider a few. These are:

1. Pie chart or pie graph (**categorical data only**).
2. Bar chart or bar graph (**categorical data only**).
3. Histogram (**quantitative data only**).
4. Box plot or box-and-whisker plot (**quantitative data only**).

Of these four, we will only show how to create the box plot by hand. We will start with that one.

Constructing a Box Plot

1. Create a "box" using Q1 and Q3.
2. Draw a line in the box at the median.
3. Draw a line, or "whisker," from Q1 to the lowest observation not considered an outlier and another whisker from Q3 to the largest observation not considered an outlier.
4. Any observations considered outliers using the 1.5*IQR approach, mark with an asterisk: *.

Example: Using the Duke men's heights, we would create box from 75 to 82; draw a line in this box at 79; make lines from 75 to 72 and from 82 to 83; and with no outliers, there are no asterisks. Figure 1.3 illustrates the box plot for this data.

Figure 1.3: Boxplot of player heights for men's Duke basketball

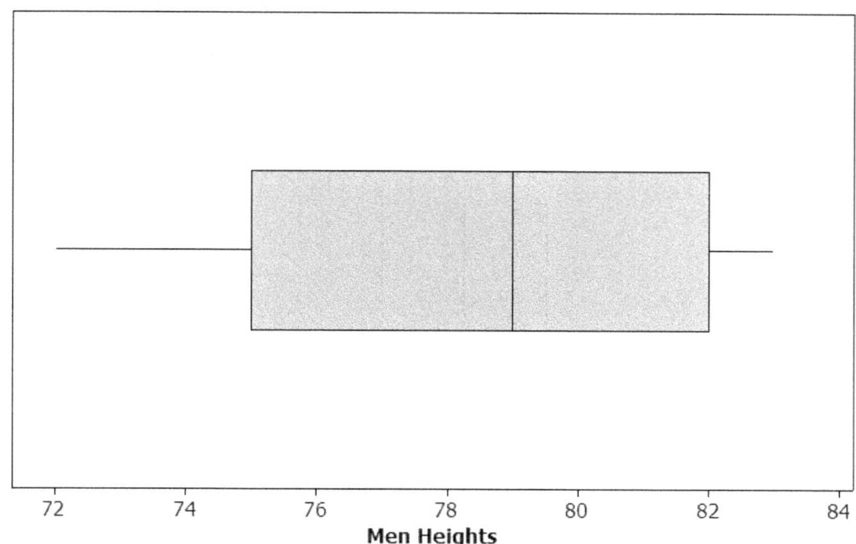

Histogram: An example of a **histogram** was provided previously in Figure 1.1 when discussing the Empirical Rule for the 2009 NFL points per game by team. That graph is repeated here in Figure 1.4, except the bell-shaped curve is removed. This curve was added specifically for that purpose to illustrate how the shape of the data was approximately bell shaped and allowed us to apply the Empirical Rule.

Figure 1.4: Histogram of NFL points per game without fit.

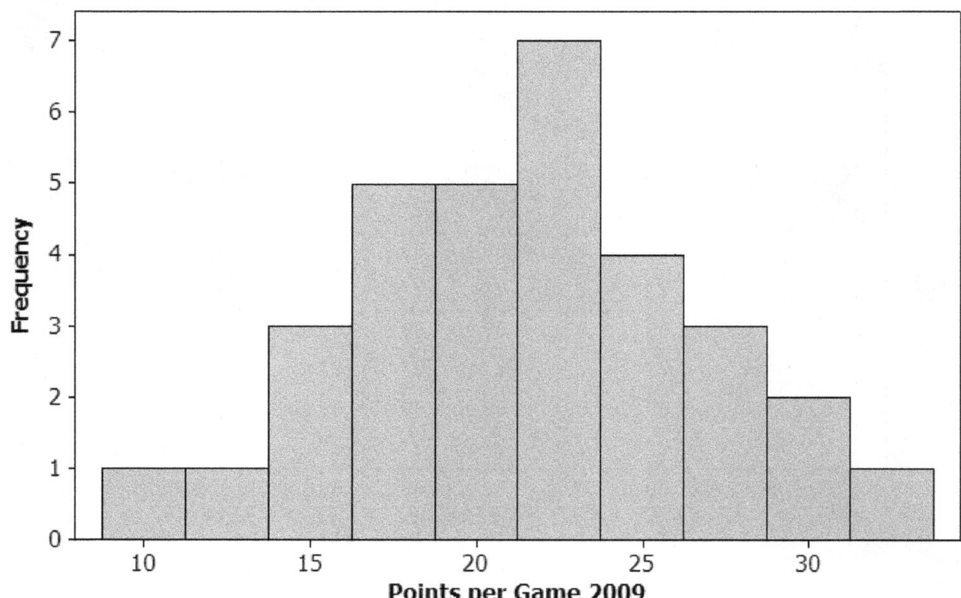

The **frequency** going up the vertical or y-axis represents a count, or the number of observations within a specific histogram bar. For instance, the tallest bar at about 23 points per game has a count of seven. This would be interpreted as seven teams average around 23 points per game during the 2009 NFL season.

Box Plot versus Histogram: Although both graphs can be used for quantitative data, an advantage of the box plot is the ease in which one can graph—and thus compare—quantitative data across categories. For example, the histogram above is for NFL points per game in 2009. What if we wanted to compare this year to points per game in 2010? The better choice would be to use a box plot for each year in a **side-by-side box plot**. From Figure 1.5, we can make quick and easy comparisons of summary measures such as median (very similar), minimum and maximum (2010 higher in both), IQR (2010 was less spread out within the middle 50% of teams), providing an overall comparison of variability (2010 appears to have less variability).

Figure 1.5: Side by side boxplot for NFL points per game in years 2009 and 2010

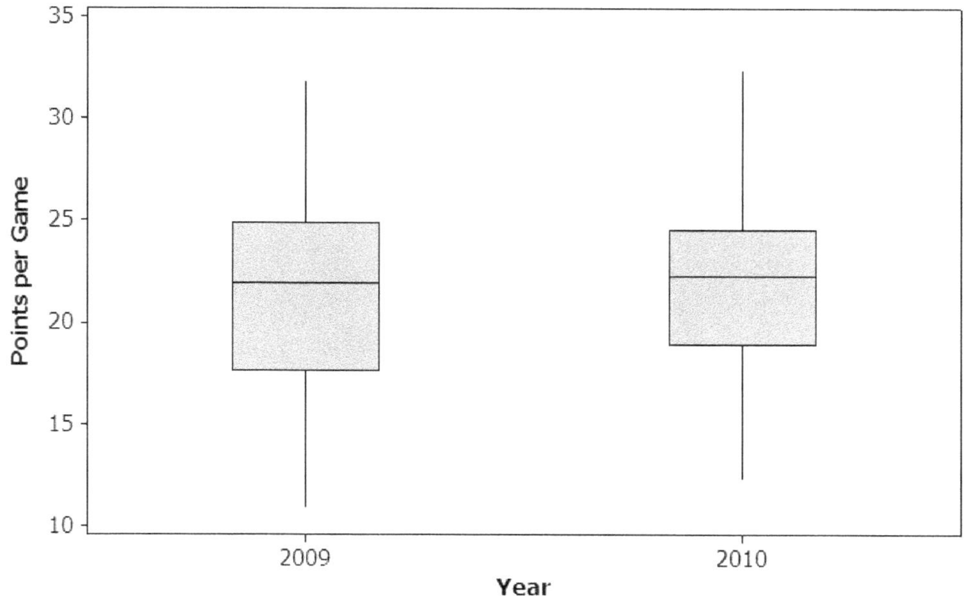

Bar and Pie Graphs: To demonstrate these two graphs, we are going to introduce a new set of data using data from 2012 men's college basketball tournament. Traditionally, there are the six major conferences (all members of the BCS) and everyone else. Our interest is in graphing the number of at-large teams receiving bids to the tournament based on conference affiliation: ACC, Big 12, Big East, Big Ten, Pac-12, SEC, or from one of these non-BCS schools. In a bar chart (see Figure 1.6), the height of the bar gives the frequency or count for each of these categories. A pie chart (see Figure 1.7) looks exactly as it sounds—like a pie. Each "slice" represents the count or percentage of observations that fall within that category.

The y-axis on the bar graph makes interpretation a bit easier, as one can quickly get a count for each category; for instance, we can see that four ACC schools received at-large bids. However, whatever software you use to create these graphs typically provides labeling options. For example, you can prompt the software to label the pie graph with counts or percents per each section.

Bar Graph versus Histogram: When starting out in statistics, one sometimes confuses the bar graph and histogram. First, remember that a bar graph is used for categorical data and the histogram is used for quantitative data. Second, the bars of the bar graph (as well as the slices of a pie graph) can be rearranged by count, size, or alphabetically because the concern is the counts within the categories and not with how they are ordered. However, since a histogram is used for quantitative data, there is a sense of order within the data. For example, 30 points per game is more than 20 points per game, and as a result the 30 should always come after the 20.

Figure 1.6: Bar graph of at-large bids by conference for 2012 March Madness

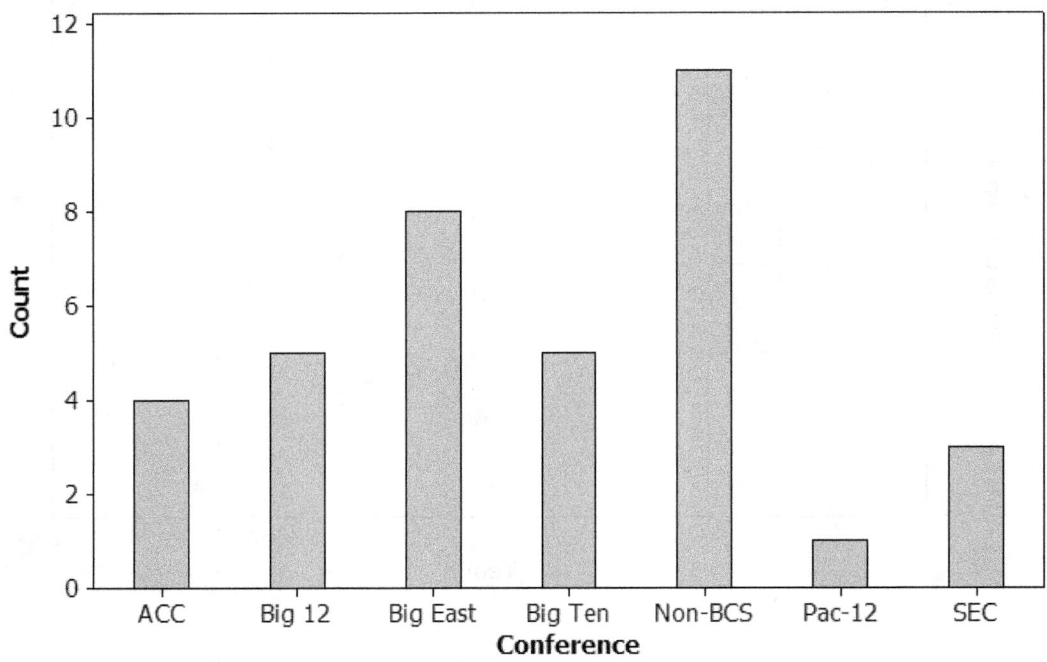

Figure 1.7: Pie chart of at-large bids by conference for 2012 March Madness

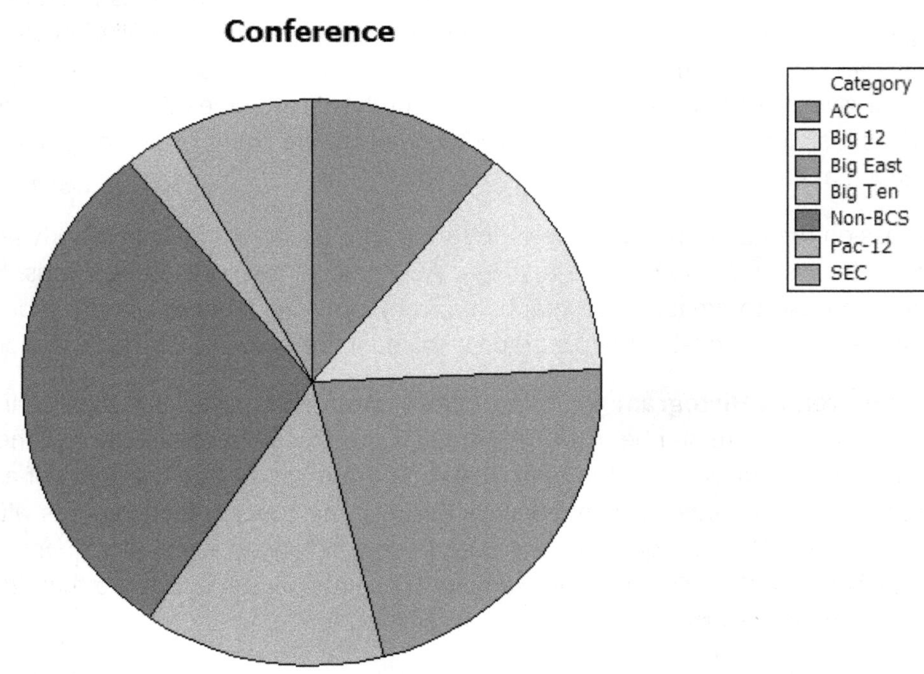

Shape or Distribution of Data—Quantitative Only!

A key aspect of *quantitative* data is the **shape** or **distribution**. Note that this concept refers only to quantitative data, as there is no concern over the shape of categorical data. Data typically follows a bell shape or a skewed shape. If **skewed**, the shape can be skewed in a negative direction (to the left) or a positive direction (to the right). This gives two classifications for skewness: negative, or left, skew and positive, or right, skew.

Importance of Shape

As previously discussed under "Better choice: mean or median," the shape of the data can have an effect on the mean and median. In that section, we stated that for skewed data the median is a better representation of the average, whereas for a bell shape, the mean is the better choice. What causes a distribution to be skewed? The presence of unusual observations. When a data set has some unusual observations on the upper end, the data gets "pulled" to the right; when these unusual observations are on the lower end, the data gets "pulled" to the left. When the data falls symmetrically around the mean, the data presents itself in a bell shape. Putting these pieces together, we have the following relationships:

Bell shape: Mean, median, and mode will all be roughly equal (see Figure 1.8).

FIGURE 1.8: Bell shape.

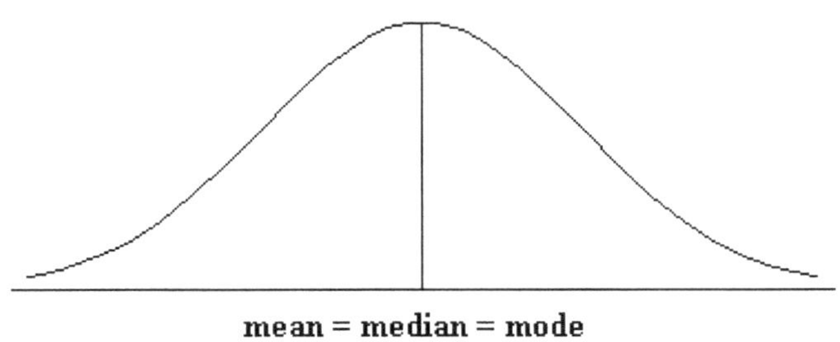

Left or negative skew: Mean will be less than the median (i.e., fall to the *left* of the median) and the median will be less than the mode (see Figure 1.9).

FIGURE 1.9: Left skewed or negatively skewed.

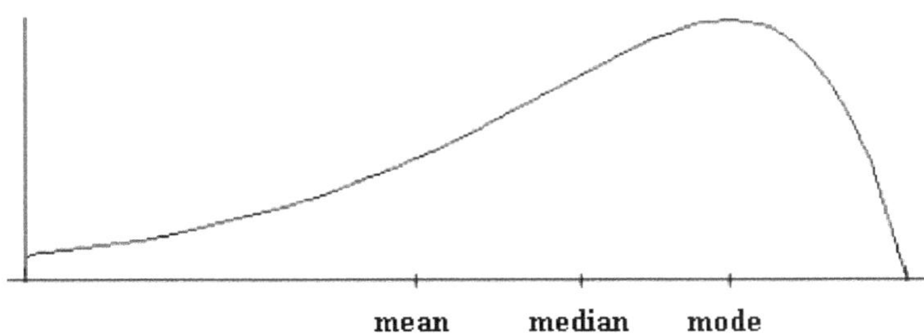

Right or positive skew: Mean will be greater than the median (i.e., fall to the *right* of the median) and the median will be more than the mode (see Figure 1.10).

FIGURE 1.10: Right skewed or positively skewed.

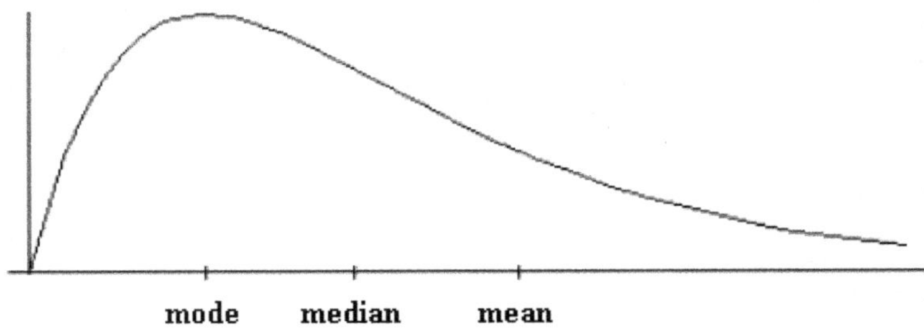

Examples: From the histogram of the 2009 NFL points per game, we could see that the data was approximately bell shaped, with a mean of 21.5 ppg. The median for this data was 22 ppg—pretty close! The histogram in Figure 1.11 provides an example of a skewed data set for NFL player salaries (in millions of dollars) for 2010. You can see that the data is clearly pulled to the right, making the shape skewed right, or positively skewed. (Observational note: Yes, that is a data point out there beyond the 24 million mark. That salary is for Albert Haynesworth from the Redskins. This salary included a 21 million dollar signing bonus!) From our discussion on the relationship between mean and median for such a shape, we stated that the mean would be greater (to the right) than the median. This is true here. The mean for this set of data is 1.7 million and the median is 0.86 million, or about $860,000. From this you can see the effect the unusual observations have on *dragging* the mean along with them.

Figure 1.11: Histogram of 2010 NFL player salaries in millions of dollars.

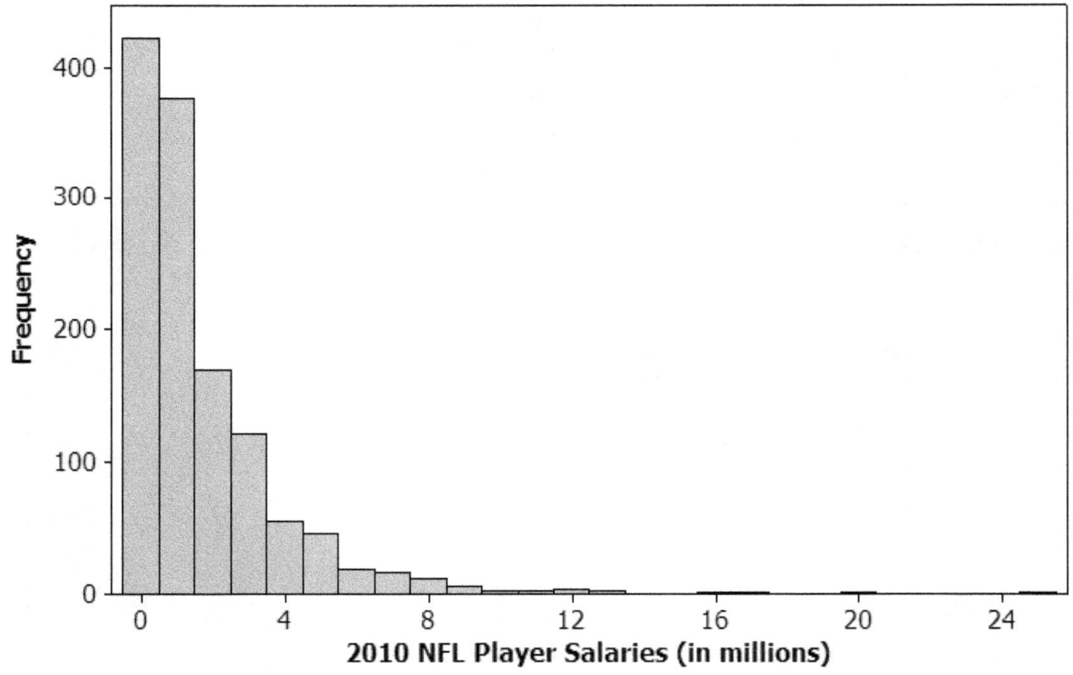

SPECIAL NOTE: This skewed right shape for NFL salaries is not unique; quite the contrary. In fact, almost all salaries, be it for professional teams or private businesses, take on a positively skewed shape. Think of your favorite pro sports team or even a place where you have worked, and consider the salaries the players or employees make. In almost every case, there are only one or two players/employees making an extreme amount. This has the tendency to skew the data to the right.

1.5 Z-Score for Bell-Shaped Data

Another useful tool in statistics when the data is bell shaped is the **z-score**. The **z-score** provides the number of standard deviations an observation is from the mean. The z-score can be used as another method to identify outliers and can be very helpful in making relative observations from two different sets of data.

Finding the Z-Score: The z-score, or z, can be found by the following, where the observed score is an observation from the data:

$$z = \frac{\text{observed score} - \text{mean}}{\text{standard deviation}}$$

Example: Returning to our Duke example, let us find the z-score for the shortest person on each team. From the data, the means and standard deviations were 78.5 and 3.9 for the men, and 72.1 and 3.6 for the women. The shortest male was 72 inches and the shortest female was 69 inches. Their respective z scores would be:

$$z = \frac{\text{observed score} - \text{mean}}{\text{standard deviation}} = \frac{72 - 78.5}{3.9} = \frac{-6.5}{3.9} = -1.67$$

$$z = \frac{\text{observed score} - \text{mean}}{\text{standard deviation}} = \frac{69 - 72.1}{3.6} = \frac{-3.1}{3.9} = -0.86$$

Interpreting Z-Scores

From our example, the male player of 72 inches is 1.67 standard deviations below (note the *negative* z-score) the mean of the 78.5 inches. The female player of 69 inches is 0.86 standard deviations below the mean of 72.1 inches. Relatively speaking, the shortest female player in comparison to the shortest male player is closer to the average height on her team than is the male player to the mean of his team. Interesting to note, both teams have a player listed as being 76 inches tall. For the male player, he would have negative z-score since his height is below the mean (considered shorter than average), while the female player would have a positive z score since her height is higher than the mean (and thus be considered a taller player than average).

Have you ever heard the phrase, "She is tall for a girl"? The z-score can be helpful to examine that question. For instance, both Duke teams have a player 76 inches tall. For the male, this translates to a z-score of - 0.64, while for the female this height results in a z-score of 1.08. The male player's height of 76 inches is 0.64 standard deviations *below* the mean, implying he is shorter than the average male player on the team. Conversely, the female player who is 76 inches tall is 1.08 standard deviations *above* the mean, implying she is taller than the average female player on the team.

Z-Scores for Outliers: We can use z-scores to identify outliers when an observation or observations fall outside three standard deviations from the mean. That is, if the calculated z-score is less than negative 3 or greater than positive 3, then that observation would be considered an outlier.

Common Mistakes: As with many of the equations in this chapter, the z-score formula appears quite simple—yet students still make several mistakes. The **most common errors** exist in the calculation of the numerator (i.e., the Observed - Mean). **One error** is that you naturally want to put the larger number first instead of substituting the correct values. For instance, in calculating the z-scores for the above male players, a frequent error is to go "78.5 - 72," because the 78.5 is larger than 72. However, by doing this, the z-score would always be positive! To avoid this problem, think of these equations as puzzles, where you just want to substitute the given numbers into their respective puzzle locations. With 78.5 being the mean, this value gets placed in the "mean" puzzle location, etc.

The **second error** results from not following the order of operation. Keep in mind that the numerator is in parentheses and must be completed first before dividing by the standard deviation. Using the male Duke height of 72, here is what we mean:

ERROR: 72 - 78.5/3.9 = 72 - 20.1 = 51.9

Note how forgetting the parentheses produces a widely different—and incorrect—z-score. So don't forget the parentheses!

CORRECT: (72 - 78.5)/3.9 = -6.5/3.9 = -1.67

Another useful application of z-scores is to compare athletes from different eras. For example, if one wanted to argue whose single season home run performance was best—Babe Ruth, Roger Maris, or Barry Bonds—one could calculate z-scores, and whichever player's z-score was highest would be considered to have had the best season (courtesy www.baseball-almanac.com). Using the Top 10 homers hit in each of the years 1927 (Ruth), 1961 (Maris), and 2001 (Bonds), we get the following z-scores:

Ruth: $z = \frac{60 - 28.5}{14.38} = \frac{31.5}{14.38} = 2.19$

Maris: $z = \frac{61 - 45.1}{7.78} = \frac{15.9}{7.78} = 2.04$

Bonds: $z = \frac{73 - 52.6}{9.59} = \frac{20.4}{9.59} = 2.13$

In comparison, we can see that each player clearly performed well above the mean of the Top 10 for his respective year, all three players hitting above two standard deviations of the mean. Notice the means and standard deviations for each year. Ruth's numbers are certainly much greater than the mean in his era, but the variability is also much larger. This has a counter-effect to the large difference in observed score versus mean. This illustrates the importance of including variability in such a discussion instead of directly comparing an observation to the mean of the observations. As to a final decision on best performance: Although Babe Ruth produced the higher z-score and therefore had the best year, one could argue that the differences are not great. This argument leads to another concept called statistical significance, where we would apply statistical methods to determine whether or not there is a statistical difference between these performances. That topic will be discussed in future chapters.[1]

Expressions and Formulas

1. $\frac{\sum_{i=1}^{n} X_i}{n}$ is the formula to calculate the mean of a sample variable "X". The symbol "$\sum$" is shorthand for "summation," meaning "to sum." The subscripts and superscripts are there to point out the starting and end points for this summation. The "$i = 1$" reflects starting at the first data point (NOT to be confused with starting with the value of 1) and continuing on to the last observation, represented by "n." The "n" also represents the size of the sample.
2. $\bar{X}$ (read "X bar") represents the sample mean.
3. The formulas for finding the locations of quartiles within an ordered string of data:

   median: $\frac{n+1}{2}$ first quartile: $\frac{n+1}{4}$ third quartile: $3 * \left(\frac{n+1}{4}\right)$
4. Variance or $S^2 = \frac{\sum_{i=1}^{n}(X_i - \bar{X})^2}{n-1}$
5. Standard Deviation or $S = \sqrt{Variance} = \sqrt{S^2} = \sqrt{\frac{\sum_{i=1}^{n}(X_i - \bar{X})^2}{n-1}}$
6. Empirical Rule, or as it is sometimes referred to as the 68-95-99.7 Rule, applies when the data is a bell shape. If bell shaped, then the observations will approximately meet the following guidelines:

   68% of the observations will fall within **1 standard deviation** of the mean:

   95% of the observations will fall within **2 standard deviations** of the mean:

   99.7%, about all, of the observations will fall within **3 standard deviations** of the mean:
7. Interquartile Range or IQR = Q3 - Q1, which is the difference between the first and third quartiles. The IQR represents the range of the middle 50% of the observations.

1 The Top 10 were used as this produced a bell-shaped distribution of home runs necessary to use the z-score method. This may have been accomplished using a larger number of players; however, in using all of the home runs hit in these respective years, the data was not bell shaped. Also, the Top 10 were from a combination of both leagues and not just the Top 10 from the players' respective leagues.

8. Find outliers using IQR and the quartiles by building a "fence" around the data. Observations outside the fence can be considered possible outliers. The lower and upper fence points are found by:

   Lower Fence: Q1 - 1.5* IQR and Upper Fence: Q3 + 1.5*IQR

9. Five number summary for a set of data consists of: minimum value, Q1, median, Q3, and maximum value.

10. Two common skews are **negative (or left skewed)**, where the mean is much less than the median. The data is being "pulled" to the left. The other skew is **positive (or right skewed)**, where the mean is much greater than the median. Here, the data is being "pulled" to the right.

11. A z-score or z-value represents the number of standard deviations an observation is from the mean for a bell-shaped set of data. Z-scores for a specified observation can be found by:

$$z = \frac{\textit{observed score - mean}}{\textit{standard deviation}}$$

Gathering Data 2

A June 2011 study conducted by *Ipsos* (www.ipsos-na.com) estimated that 13% of U.S. teens and adults (12 years old and up) competed in fantasy sports during 2010. They break this down further to state that 19% of males and 5% of females participated in fantasy sports during that time. Impressive numbers for sure, but are they useful? That is, how do they answer the following questions:

1. How was the study conducted? Was the study designed to provide a representative sample of the population of U.S. teens and adults? How about males and females within this group?
2. Is there any expectation of error, and if so, how big (or small) of an error can be assumed?
3. Do any potential biases exist in the study?

These questions raise some new concepts to understand: *population*, *representative sample*, *error*, and *bias*.

2.1 Terminology

Population: The population is the "big picture"—or in other words, the large group of subjects to whom a researcher wants to reference his/her results. This population does not have to consist of people; it can refer to creatures (e.g., imagine estimating the number of flies buzzing around a pitcher's head during a Cleveland Indians game on a muggy September evening) or nonliving objects, for instance, how many bats produced by Louisville Slugger will be rejects. In the above *Ipsos* study, the population was all U.S. teens and adults.

Sample Frame: A sample frame consists of all subjects from which the researcher will take a sample. Ideally, the sample frame and population should be identical. The sample frame may consist of all those in the population of interest that have land line phones. This, however, would be limiting, as nowadays many people no longer use land lines; they just use cell phones. In such a case, the sample frame may not be the same as the population and the study may result in a poor (think not accurate) estimate. For the *Ipsos* study, the company claims to have taken their sample from the national general population and uses several tools (e.g., phone, mail) to elicit responses.

Sample: The sample is the subset of the population taken from the sample frame. To be considered a *representative sample*, the sample must be taken randomly.

Margin of Error: Since studies are done using samples, by design they cannot fully address the entire population. Therefore, error in the estimates will exist. For example, if *Ipsos* were to conduct the study starting completely over by first getting another random sample, do you believe this second sample would consist of the exact same subjects as that in the original sample? Of course not; the two samples would most likely differ. Because of this, the percentage of the sample who say they participated in fantasy sports in 2010 may be some number besides 13%. For argument's sake, let's say this repeated study estimates 12%. Which estimate is correct: 13% or 12%? They both cannot be right—in fact they could both be wrong!—as there is a fixed percentage of the population who participated. That is, the answer may not be 12 or 13 percent but some other percentage. So how does one account for error in such a study? One method is to use a conservative margin of error, which can be found by taking $\frac{1}{\sqrt{n}}$, where "n" is the sample size. In the *Ipsos* study, 1215 U.S. teens and adults were surveyed. This comes to an approximate margin of error of $\frac{1}{\sqrt{1215}} = 0.028$ or 2.8%.

Bias: When a sample is not representative of the population, the results are said to contain **bias**. There are several causes as to why bias may occur. Bias may occur due to: *sampling*, *response*, or *non-response*. More on these later.

Response Variable: This is the outcome variable; e.g., does one participate in fantasy sports?

Explanatory Variable: A variable is used to explain why some outcome occurs. For instance, maybe one's participation in fantasy sports is related to whether or not the person has another family member participating in fantasy sports.

Confounding Variables: These are explanatory variables that are associated with both the response variable and each other.

Lurking Variable: A lurking variable is any variable not included in a study that could affect study results. These differ from confounding variables, as lurking variables were not part of the study; they *could be* confounding variables if included.

2.2 Methods to Gathering Data

In general, there are two ways to gather data, i.e. take a sample: non-probability methods or probability methods. The second method is preferred, as this format allows one to assume the sample to be representative of the population from which the sample was taken.

Non-Probability Methods: Convenience Sampling and Volunteer Sampling

Although there are several such techniques, two of the most common methods are convenience and volunteer sampling.

Convenience: Have you ever been stopped on campus or on your way to work/school and asked to participate in a survey? This is an example of convenience sampling, as the surveyor can gather the data easily and cheaply: They can just stand at a popular spot and interview those who pass by.

Volunteer: This the most common convenience sample, where subjects are asked to *volunteer* to participate. Anytime you watch ESPN or go to ESPN's or yahoo sports website, for example, you are often requested to answer a question on who is going to win the Super Bowl or the BCS game, etc. These are examples of volunteer sampling. The name "volunteer" comes from you being asked to volunteer your input.

These sampling methods are not ideal, since they typically do not reflect the population of interest. For example, you may never walk by that popular campus spot or refuse to partake in the online surveys. Also, in the case of online surveys, there are possibly no controls on how often one can participate. That is, a person or group could repeatedly offer their input on which team is going to win.

Probability Methods: Simple Random, Stratified Random, and Cluster Sampling

A **simple random sample**, or **SRS**, is the one you are most likely familiar with and the easiest to understand. As the name implies, a sample is taken randomly from the population and is sometimes referred to as a **random sample**. The main idea of a random sample is where each subject in the population has an equal chance of being selected.

Stratified random sampling begins by dividing the population into specific groups called **strata**, then taking a random sample from each strata. The purpose of this method is to guarantee participation in the study of specific groups (e.g., gender) in which the researcher may believe a difference in attitude, behavior, etc., may exist. For example, one may believe that males are more likely to participate in fantasy sports than females. Therefore, when conducting a study to estimate participation, you would want to divide your population by gender (i.e., stratify your population into males and females), then sample randomly from each group; that is, take a random sample of females and a random sample of males. This will guarantee that your overall sample will include members from specific, key groups. From the prior *Ipsos* survey, the sample was weighted to balance demographics and ensure that the general population sample's composition reflects that of the U.S. age 12 and older population according to recent U.S. Census data. The implication is that the population was stratified and the sample was taken so that the percentage of females and males in the sample was reflective of the percentage of females and males who make up the 12 and over population.

With **cluster random sampling**, the population is divided into "clusters," and then these clusters are selected randomly. Each subject within each randomly selected cluster makes up the sample. For example, say you wanted to conduct a survey to estimate how many students at your college or university participate in fantasy sports. You could list all of the campus dorms and downtown student apartment complexes as clusters, then randomly select from this list several dorms and apartments. After this, each student living in those buildings selected would be interviewed. Notice that you did not take a random sample from each dorm and apartment but instead randomly selected dorms and apartments.

SPECIAL NOTE: Consider this cluster sample: If you felt there might be a difference between college students attending sporting events based on living on campus versus off campus, you could first stratify the buildings by classifying them as either on or off campus, then take a cluster sample from each of these strata. This would be classified as stratified cluster sampling, since the method includes both the stratified and cluster sampling techniques.

Common Mistake

A common mistake is often made between stratified and cluster sampling. The difference is that in stratified sampling, a random sample is taken from each strata. For instance, if the population is divided into females and males, a random sample is taken from each group. This results in the final sample consisting of females and males. However, in cluster sampling, it is the clusters that are randomly sampled. Again, if you divided the population into females and males and considered these clusters, then you would randomly sample from these clusters (e.g., randomly select one of these two clusters), and thus all members of that selected cluster would make up the sample. For instance, if you randomly selected the cluster "male," then all the males would comprise the sample. You can see how this would result in a completely different sample from the stratified sample, which would produce a sample including both sexes.

2.3 Types of Bias

As mentioned previously, bias can result when the sample is not representative of the population, or by how the subjects respond—or don't respond—to the survey.

Sampling Bias: When the sampling method does not reflect the population, this can produce bias. Nonrandom sampling methods are prone to bias, as are sampling methods that produce a sample that does not properly represent specific groups. For instance, instead of using a random sample method, say the *Ipsos* survey opted to visit various high schools and colleges located in metropolitan areas. The resulting sample size may have been large, but the sample subjects may not be representative of the general population. Or what if care wasn't taken to get a sample that represented the percentage of males and females in the population? The study results may have been biased by having too many males (or females) participating. This is an example of *undercoverage* of a sample. Random sampling methods are designed to provide representative samples, but if these are not conducted properly (see the last example) they, too, can produce biased results.

Response Bias: As the name implies, the bias is due to how subjects respond. The bias results because the subjects do not reply correctly or honestly. Think of yourself sitting among a fellow group of students and someone asks you if you like the Steelers. Well, maybe you hate the Steelers and all the annoying behaviors of Steeler fans (e.g., the waving of the terrible towel and the raising of six fingers), but you also know everyone around you is a Steelers fan. Instead of giving an honest response, you respond by saying you are a fan, and thus avoiding a verbal whiplashing! This type of bias can also result when the question asked is misleading or poorly worded. For instance, imagine again you are sitting in this group of Steeler fans and you are asked, "So, I hear you hate the Steelers. Is this true?" Being asked a question in this way may prompt you to refute this statement, and instead you reply that you don't hate them.

Non-Response Bias: Similar to response bias, the title gives insight to the definition. Whereas with response bias the subject still offered a response, with non-response bias the subjects fail to respond. This can occur from subjects in the sample not being reached or if reached, they refuse to answer.

2.4 Types of Studies: Observational versus Experiment

An **observational study** is one where the subjects are observed on their behavior, while in an **experiment**, there is a random assignment of some treatment. If given a choice, a researcher should prefer the experiment, as such a method can allow one to reach **cause and effect** or **causal** conclusions. A *causal* conclusion is one where a treatment can be said to *cause* an outcome. This result cannot be reached with an observation study. In an observational study, a researcher could only conclude that a *relationship* or an *association* exists between some response and explanatory variable.

Observational Studies

Questions to consider:

1. Do the Saints play better (i.e., win more) when they play indoors versus outside? The response variable being game outcome (win/lose) and the explanatory variable being location (inside/outside).
2. Does LeBron James's free throw percentage on his second attempt depend on how he does on his first attempt? The response variable being second attempt (make/miss) and the explanatory variable being first attempt (make/miss).
3. If an NFL player suffers a concussion, will this result in memory loss? The response variable being memory loss (yes/no) and the explanatory variable being concussion (yes/no).

For each question, there would be no randomization. Therefore, one could not conclude that playing outdoors *caused* the Saints to lose, or missing his first free throw did not *cause* LeBron to miss his second attempt, or having a concussion *causes* memory loss.

So why even conduct an observational study? As these examples illustrate, there can be various reasons. How could you expect LeBron to randomly miss and make his first attempt, or how could you ethically instruct an NFL player to suffer a concussion? For such reasons, an experiment is not always possible.

REMEMBER: Association does not mean causation. Just because one might find a relationship between a set of variables, this shouldn't imply that one of the variables caused some outcome to occur.

Experiments

Questions to consider:

1. Can the energy drink *5 Hour Energy* improve a player's energy level? The response variable being improved energy level (yes/no) and the explanatory variable being the taking of *5 Hour Energy* (yes/no).
2. Which golf ball is longest? The response variable being distance and the explanatory variable being golf ball.
3. Would the use of tennis balls for practice improve a receiver's skill in catching a football? The response variable being improve skill (percentage of balls caught) and the explanatory variable being use of tennis balls (yes/no).

For each of the above, one could perform an experiment where the explanatory variable is randomly assigned to the subjects. In an experiment, this explanatory variable is referred to as a **treatment**.

Designing a Good Experiment

The design of a good experiment includes two key components:

1. A control group.
2. Randomization.

A **control group** is used to allow for a comparative result. What sense would a study make if all subjects were given the energy drink? How would a researcher know if the drink had any effect on energy level if there was no result from a group not taking the drink? If you kept hitting the same golf ball, how would you know if another ball would go further (or shorter)? In order to make such decisions, one must have a control group. As these examples demonstrate, a control group is not always a group that doesn't receive a treatment. For instance, in the golf ball example, you are still hitting a golf ball; however, you need to have at least two groups in order to make a comparison.

Often, a **placebo** will be used in an experiment. Think of a *placebo* as a treatment without value. For instance, if conducting an energy drink experiment, one may want to assign subjects to a non-energy drink that tastes and looks like the energy drink. If a placebo is included, note the possibility of a **placebo effect**. This phenomenon occurs when those taking the placebo may respond better compared to those receiving nothing. You might be familiar with the common term for a placebo, *sugar pill*.

With **randomization**, the experimenter randomly assigns treatments to the subjects, or conversely can randomly assign subjects to treatments. Commonly, subjects are randomly assigned to treatments. If the researcher allows the subject to pick the treatment, we are back to using an observational study. The focus of randomization is to:

1. Eliminate bias if subjects were allowed to choose or if the researcher was allowed to assign. For instance, the more natural receivers may want to try the new technique, and therefore more of this athlete selects this treatment, or conversely, the researcher might want to bias the groups to improve chances of success (i.e., conclude a favorable result). In either event, the result can be unbalanced groups (see 2 below).

2. Balance the groups (e.g., you would want naturally gifted receivers in both the use and nonuse of tennis balls).

3. Balance out any effects due to lurking variables.

Understandably, all treatment groups must be treated equally. To aid in this effort, a researcher should consider **blinding the study**. By blinding the subjects, the researcher reduces the chance of the placebo effect, as the subject would not know whether they were receiving the treatment (energy drink) or the placebo. Typically, the researcher is interested in a positive outcome, an outcome that illustrates a cause and effect. They, too, could be biased in how they gather the results. They could word questions in a way that provokes a more positive (or negative) response on the subject based on which group—treatment or placebo—the subject belongs. To avoid this consequence, the researcher and those administering the treatments should be blinded as well. When neither the subject nor those involved with the subjects are aware of which treatments are being assigned, the study is referred to as a **double blind** study. This type of study is ideal.

Example Conducting Experiment: Energy Drink

First, we will assume there are no serious side effects associated with *5 Hour Energy* to avoid use of randomization for ethical reasons. To start, we would need to gather a group of willing subjects, for instance the Steelers.

Since we want to measure a change in energy level, we would need to have some method of measuring energy level and then apply this to each player to get a starting point called a **baseline**. This baseline allows us to see if any improvement is gained following treatment. One possible measuring device could be a questionnaire. The players are administered this questionnaire and scores are tallied based on responses. This is their baseline energy level.

Next comes the randomization. Using a placebo—a drink similar in taste, odor, color, etc., of the energy drink—and the energy drink itself, players are randomly assigned to receive either the placebo or energy drink. Players and treatment administrators are blinded as to who gets what. With this randomization, we expect to have control over any biases, such as one group having better-conditioned athletes. We also hope to control for any lurking variables, such as hours of sleep the night before, whether or not the player recently played in a game, or minor injuries. The goal is that both groups, placebo and treatment, have a similar number of players for whom such lurking variables exist.

After taking their treatment, the players are again offered the questionnaire and the scores are compared to their baseline scores where increased scores are indicators of higher energy levels. If the energy drink improves energy levels, you would expect the group receiving the drink to have increased scores compared to the scores from the placebo group. If this were the case, one could conclude that *5 Hour Energy* causes an increase in energy level—but to what degree?

Generalizing Results: Random Selection versus Random Assignment

For the most part, people understand the concept of *generalization*, where study results are said to be indicative of some larger group: the population. However, confusion enters on *when* this can be done. This is the difference between random selection and random assignment.

Random selection refers to the method in which the sample was taken. If a study is done using random sampling methods, then those results can be extended to that population from which the random sample was taken. This is true regardless of study type, observational or experimental.

However, **random assignment** relates to an experiment and whether or not one can state *cause-and-effect* conclusions. If random assignment is used, then cause can be concluded.

As stated in the Steelers *5 Hour Energy* example, since random assignment was used, an increase in energy level could be said to be caused by the use of *5 Hour Energy*, but to what population could this result be generalized? If the Steelers were selected for convenient reasons—the researcher worked for the team, for example—then the results of the experiment can only be made to the Steelers. However, if the Steelers were randomly selected from a list of all 32 NFL teams, then one could generalize the results to all NFL players. This is the difference—and importance—of random selection and random assignment. A **complete randomized experiment** is an experiment incorporating random selection and assignment. Of all possible study designs this one is best.

2.5 Practice Examples

Example 1: College bowl season has arrived, and ESPN is interested in getting people's opinions on who will win each game. ESPN the website and ESPN TV and radio broadcasts encourage viewers and listeners to visit the ESPN website to make their picks. Just prior to the start of the games, ESPN publishes the poll results saying, "This is how American sports fans voted on who will win each bowl game." From this, answer the following:

Question 1: Based on how ESPN reported the results, who does ESPN believe to be their population of interest?

Question 2: What type of sampling method was used to get the data?

Question 3: To what population do the poll results actually represent?

Answer 1: From ESPN's statement, "This is how American sports fan voted," the organization is implying that the results reflect all sports fans in America. A fairly ridiculous conclusion, don't you think?

Answer 2: The method used was a convenience sampling, specifically a volunteer sample, since you voluntarily went to the website to cast your vote.

Answer 3: Since no random sampling was done, the results just reflect those who voted for *each game*! For example, the results for the Rose Bowl only reflect the opinion of those individuals that cast a vote for the Rose Bowl; the results for the BCS Championship only reflect the opinion of those who casted a vote for the BCS Championship, etc. The results do NOT reflect the opinions of all the viewers, those who knew about the poll but didn't vote, or even all those who visited the website. The results, again, can only be thought of as representing those individuals who actually took part in the survey.

Example 2: You are an Ohio State fan and want to estimate the proportion of OSU fans who are excited about the hiring of Urban Meyer. Answer Question 1 one below, plus provide a sampling method that would represent each of the 2 through 5 (multiple correct answers are possible).

1. From the example, what is the population of interest, and how would you go about getting the people that make up this population?
2. Selection bias.
3. Simple random sample.
4. Stratified random sample.
5. Cluster sample.

Answer 1: The population of interest is pretty simple: All OSU fans. However, determining how to properly list this population so one could take a sample provides some difficulty. One could simply use football season ticket holders, but then someone could argue that there are possibly more OSU fans that don't have season tickets compared to those who do have season tickets. Another option could be to use all Ohio residents, but then not all Ohio residents are Buckeye supporters. As you might realize, the question seems simple enough—basically, you want to know how OSU fans feel about their new coach—but defining that population correctly in order to get a representative sample is not so simple! A quick change to the research question, however, could make life easier: If you just stated you wanted to gauge the attitude of football season ticket holders on the new coach's hiring, then your population can be better represented through random sampling.

Answer 2: An example of selection bias would be a sampling method that included subjects who are NOT Ohio State fans. As you might imagine, Michigan fans are possibly not excited to have Coach Meyer working at their rival. If you took, for example, a random sample from all adults living in Ohio, your sample could include individuals who are not OSU fans.

Answer 3: The previous answer provides one simple random sample technique, although the method could involve selection bias. This illustrates that even though your sampling method employs probability sampling, doing so does not necessarily guard from bias. Remember that the goal of probability sampling methods is to obtain a *representative sample of the population of interest*. Perhaps a way to do this would be to take a random sample from all OSU season ticket holders, or if taking a random sample of all Ohio adult residents, you could ask if they were OSU fans; then you could throw out all those who said no.

Answer 4: For this, let's assume our population of interest is "OSU football season ticket holders." A stratified method requires that we first divide the population into some groupings called strata. This could be, for example, student versus non-student; or taking the non-student a step further and splitting into the years one has held season tickets; a third possibility could be to divide the stadium into various sections. In either case, after creating the strata, one would take a random sample from each strata.

Answer 5: A quick solution to cluster sampling would be to randomly select a row from each section of the stadium, then interview each subject in those rows. From this method, notice that the subjects themselves were not randomly selected; instead, they were selected in a group or cluster (the row). This could have been done as well with season ticket holders: Divide the ticket holders into their sections, then randomly select rows from these sections, and then interview the ticket holders in those selected rows.

Probability 3

The discussion of probability in elementary statistics can often present much confusion for students. The topic tends to contain some of the more difficult concepts for students to grasp. They either understand the concept but not the formula, or vice versa. Fortunately, there exists in sports many examples that can readily explain these concepts and provide a foundation for their understanding. For instance, consider the following two conversations between two friends:

Conversation One

Friend One: What chance do you give New England in beating the Giants in the Super Bowl?

Friend Two: I'd say it's about 50/50.

Friend One: What if Brady can't play?

Friend Two: Then New England has no chance.

Conversation Two

Friend One: What chance do you give New England in beating the Giants in the Super Bowl?

Friend Two: I'd say it's about 50/50.

Friend One: What if their second string quarterback, Brian Hoyer, gets hurt and can't play?

Friend Two: So what, that won't matter. I still give them a 50/50 shot.

If you can appreciate both conversations, then you already have a basic grasp of probability concepts such as **subjective probability**, **conditional probability**, and **independent events**.

3.1 Defining Probabilities

The idea of **probability** is simply the chance that some event occurs (or doesn't occur). From a numeric standpoint, this implies that an event either has no chance of happening (probability of 0), will absolutely occur (probability of 1), or somewhere in between. Thus, the probability of any event taking place ranges from 0 to 1. This is easy enough, and to most of us it is common sense, but *how* the probability of an event is found is another issue.

In general, there are three common types of probabilities: **Classical**, **relative frequency**, and **subjective**. *Classical probability* relates to the number of times an outcome can occur out of the total number of possible outcomes, assuming all outcomes are equally likely. For those who enjoy a bit of gambling, think of flipping a coin, rolling a die, or choosing a card from a standard deck of cards. Relatively, the probability of a head is 1/2, the probability of getting a "1" is 1/6, and the probability of getting the ace of spades is 1/52. Or, if you believe in the parity of the NFL, each team has a 1/32 chance of winning the Super Bowl before the season starts! *Relative frequency probability* is defined as what occurs in the long run or over a long number of trials. Consider two evenly matched teams playing a game; you would give each team a 1/2, or 0.5, probability of winning the game. If these teams played twice with Team A, winning both times, does this necessarily change the probability? No. This is the idea of "over a long-run number of trials." The idea is that if the two teams played over and over and over again, the numbers would eventually even out to where both teams won an equal number of times. Applying this to the classical probability examples, think of rolling the die. You may roll the die 10 times and never get a 1. Assuming the die is fair, if you rolled the die many, many times you would expect about 1/6 of these many roles to produce a 1. This raises another concept called the **law of large numbers**. *Subjective* probability is when one sets the probability of some outcome on personal views. This is sometimes referred to as *personal probability*—the belief a person gives to an outcome happening. This is illustrated in the above conversations. Friend Two is giving the Patriots a 50/50 chance of winning. There is no long-run series of trials on which to base this probability—they have only played once before in the Super Bowl (2008)—and the teams playing in 2012 are not exactly the same.

3.2 Terminology, Notation, and Math Rules

Terminology

Sample Space: The set of all possible outcomes.

Event: A subset of the sample space.

Disjoint Events or Mutually Exclusive Events: These are events that do not have any outcomes in common. That is, they do not share any outcomes: If an outcome occurs in one event. this outcome cannot occur in another event. By rule, if events are disjoint they cannot also be independent. That is, if events are disjoint, they are also dependent.

Independent Events: Events are independent when one event does not "influence" the chance another event occurs. (See Conversation Two at the beginning of the chapter.)

Complement Events: These are all events not in the event of interest. By rule, complementary events are also disjoint events.

Intersection of Events: This consists of all outcomes shared by events. Unlike disjoint events, intersection represents the outcomes that events have in common.

Union of Events: The combined outcome of two or more events.

Conditional Probability: When the occurrence of one event influences the probability that another event occurs. (See Conversation One at the beginning of the chapter.)

Simple Events: These are events that have only one outcome.

Notation

With probability, events are often defined by a long phrase. For instance, we might be interested in the event "Giants win Super Bowl" or "Giants win Super Bowl and Manning throws for over 300 yards." To simplify things, single letters are given to represent these wordy events. The usual choice of lettering is "A" and "B" when discussing two events, and then "C" if three events, etc. What is important, however, is to not let this labeling confuse you: just think of it as shorthand. For instance, one might say, "Let A be the event the Giants win the Super Bowl." However, one could just as easily say, "Let G be the event the Giants win the Super Bowl." Other than using A or G, the meaning is the same: We are interested in the outcome, Giants win the Super Bowl. The important idea is that you identify what the lettering represents, as once you identify the events with simple labels, you use the labels!

P(A): The probability that event A occurs. Remember, by earlier definition of probability, the probability of any event ranges from 0 to 1, or $0 \leq P(A) \leq 1$.

P(A and B) or P(A ∩ B): The probability that events A and B occur. This is the notation form for the intersection of two events. Think of the symbol ∩ as an "A" without the connecting line. This is also referred to as the **joint probability**.

P(A or B) or P(A ∪ B): The probability that events A or B occur. This is the notation form for the union of two events. Think of the symbol ∪ as representing union.

P(A') or P(A^c): The probability of the complement to event A.

P(A|B): The probability event A occurs *given* event B has occurred. The symbol "|" should not be confused with "divide by." That is, do not think of this as "A divided by B." This notation represents the conditional probability of "event A occurs given event B occurred."

P(B|A): The probability event B occurs *given* event A has occurred. As with the prior notation, the symbol "|" should not be confused with "divide by." That is, do not think of this as "B divided by A." This notation represents the conditional probability of "event B occurs given event A occurred."

Probability Rules

Complement Rule: $P(A) + P(A^c) = 1$ or equivalently, $P(A) = 1 - P(A^c)$. In words, the probability of an event plus the probability of the complement to that event equals one.

Independence Rule: If two events, say event A and event B, (remember, don't get hung up on notation, as this could be any lettering used!) are independent events, then the P(A and B) = P(A)*P(B). In words, if two events are independent, then the probability of their intersection is equal to the product of the individual event probabilities.

Addition Rule for Union: P(A or B) = P(A) + P(B) - P(A and B). In other words, the probability of the union of two events is found by adding the individual event probabilities, then subtracting the joint probability (i.e., the probability of the intersection of the two events).

Conditional Probability: $P(A \mid B) = \frac{P(A \text{ and } B)}{P(B)}$ and conversely, $P(B \mid A) = \frac{P(A \text{ and } B)}{P(A)}$. Notice that in both equations, the numerator is the probability of the intersection of the two events. The denominator is the probability of the given event.

Two Special Rules:

1. If two events, A and B, are disjoint, or mutually exclusive, then the P(A and B) is zero. This makes sense by the definition of disjoint, which says two events are disjoint if they do not share any outcomes in common. Thus, if two events are disjoint, the intersection of the events must be zero, as they would not have any outcomes in common. So if two events are disjoint, P(A or B) = P(A) + P(B). Note this only applies when we know the events are disjoint; otherwise use the addition rule for unions. Remember, disjoint and mutually exclusive mean the same thing.

2. If two events, A and B, are independent then P(A|B) = P(A) and P(B|A) = P(B). This, too, should make sense when you consider the meaning of independence. Recall that two events are independent when the occurrence of one event does not affect the chance another event occurs. Therefore, if A and B are independent events, then knowing event B occurred would not change the probability event A occurs, i.e., P(A|B) = P(A). This same logic applies to P(B|A). For those who prefer to see this mathematically, consider the following, keeping in mind the **independence rule** that says P(A and B) = P(A)*P(B). Applying this rule to two independent events:

$$P(A \mid B) = \frac{P(A \text{ and } B)}{P(B)} = \frac{P(A) * P(B)}{P(B)} = \frac{P(A) * P(B)}{P(B)} = P(A)$$

[Same math logic for P(B|A)]

The key to Special Rule 2 is to remember that this ONLY applies when the events are independent. If independence is not known or is not true, then you need to use the conditional probability rules.

Rules for Independent Events

If asked to show if two events are independent, one only needs to verify if any *ONE* of the following is true. If one of the following is true, then so are the other two, and the events would be labeled as independent.

1. P(A and B) = P(A)*P(B)
2. P(A|B) = P(A)
3. P(B|A) = P(B)

3.3 Applying Probability Rules: Some Examples

Probabilities are generally provided in one of two ways: You are either given the probability (e.g., the probability Team A beats Team B is 0.25) or you can calculate the probability from a series of outcomes displayed in a table. We will look at both. NOTE: These examples use probability based on relative frequency, since it is based on what has transpired over time.

Example 1: World Series Game 7

Since the World Series went to a best-of-7 format in 1905 through 2011, there have been 36 fall classics that extended the full seven games. Below is a breakdown of home team (AL or NL) and whether that home team won game 7. This particular type of table is called a **two-by-two table**, as the table is composed of two rows of data and two columns of data *excluding the rows and columns for totals*!

Home	Lost	Won	Total
AL	10	10	20
NL	7	9	16
Total	17	19	36

TABLE 3.1: Two-by-two table of Game 7.

Question 1: What is the probability that the home team was from the American League (AL)?

Question 2: What is the probability that the home team won game 7?

Question 3: What is the probability the home team won *given* the home team was from the American League?

Question 4: Are the events "Home team AL" and "Home Team Won" independent events?

Answer 1: The probability the home team was from the AL can be found by the number of total times the AL was home team (20) divided by the total number of games (36). The answer is therefore 20/36 = 0.555.

Answer 2: The probability the home team won can be found by the number of total times the home team won (19) divided by the total number of games (36). The answer is therefore 19/36 = 0.528.

Answer 3: This is a *conditional probability* and we start with the *given* event, which is the home team was from the AL. From the table, there were 20 instances where the home team was from the AL. Then, of these 20, the home team won 10 times. The probability could be written as follows: P(Home Wins | Home AL) = 10/20 = 0.50.

Answer 4: Recall that to state independence, we need to show that *one* of the following is true:

$$P(A)*P(B) = P(A \text{ and } B) \text{ or } P(A|B) = P(A) \text{ or } P(B|A) = P(B)$$

From the table, we can calculate the probability that "Home Team AL" and "Home Team Wins" from taking the number of times these two events occur together (10) divided by the total number of games (36). So the P(Home Team AL and Home Team Wins) = 10/36 = 0.28.

Starting with using some notation, let's define A as "Home Team AL" and B as "Home Team Wins." Applying this to the above formulas:

P(A and B) = 0.28 and from answers to Questions 1 and 2, P(A)*P(B) = 0.555*0.528 = 0.29.

Pretty close! And this presents one possible problem: rounding! If we had rounded to more (or fewer) decimal places, the answer may have been more clear. In this case, one could reasonably conclude that these are "close enough" to assume independent. However, technically speaking, since they are not *exactly* the same, one might say they are not independent. Fortunately, statistics can help provide us with an answer to this question of, "Are they close enough?" which is a topic we will discuss starting in Chapter 6.

From Answer 3, we have P(B|A) = 0.50 and from Answer 2, we have P(B) = 0.528. Again, pretty close—which it should be, as the decision on independence should be the same regardless of which equation you use. Note that the final probability may not be the same—that is, above we are comparing 0.28 to 0.29 and here we are comparing 0.50 to 0.528—but the decision you make will be the same. So once more, like before, possibly too close to call!

What one can conclude from this information is that, in general, there is no advantage to playing game 7 at home. However, a possible lurking variable (remember these from Chapter 2!) could be the weather or starting pitcher. For example, maybe northern teams are more likely to win game 7 at home compared to if the game were played in the South, as the October/November weather in the North could be a factor. Perhaps a team's best pitcher had to be used in game 6, leaving that team to play against the other team's ace.

Example 2: Predicting Super Bowl Winners

Some say, "Offense is what makes the game exciting," but does this translate to winning the Super Bowl? From reviewing the regular season points per game (ppg) averages for teams playing in the Super Bowl (see www.nfl.com), the probability is 0.57 that the team with the higher ppg average during the regular season wins the Super Bowl. The probability is 0.52 that the team with the higher ppg is from the American Football Conference (AFC). Finally, the probability is 0.27 that the team with the higher ppg is from the AFC and wins the Super Bowl.

Question 1: What is the probability that the team with the higher ppg *loses* the Super Bowl?

Question 2: Are the events "Higher ppg wins Super Bowl" and "Team with higher ppg is from AFC" independent?

Answer 1: This question requires that we look at the *complement* of an event. Since the team can either win or lose the Super Bowl (disjoint events since a team cannot both win *and* lose the same game!), the probability the team with the higher ppg loses the Super Bowl can be found by:

1 - P(Team with higher ppg wins Super Bowl) = 1 - 0.57 = 0.43

Answer 2: Again, for independence, think of using any one of the above mentioned formulas in Example 1. Since we are given a "joint probability" (remember these are "and" probabilities), we can use the P(A and B) = P(A)*P(B) to check for independence.

If we define A as "Team with higher ppg wins" and B as "Team with higher ppg is from AFC," then from the example we have:

$$P(A) = 0.57,\ P(B) = 0.52 \text{ and } P(A \text{ and } B) = 0.27$$

Plugging these probabilities into our equation P(A)*P(B) we have:

(0.57)*(0.52) = 0.29, which is again close to our P(A and B) of 0.27! You can see how frustrating probability can be, as the results are not always so clear!

From this information, it would appear that having the higher scoring offense in the regular season does not translate into an advantage in the Super Bowl. We leave it to you to research www.nfl.com or some other resource to see if defense would play a role; that is, is the team with the lower ppg allowed during the regular season more likely to win the Super Bowl?

Example 3: Is There a Home Field Advantage in NFL Championship Games?

There have been 46 Super Bowls through 2012, meaning there have been 92 conference championships: 46 in each league. For each of those championship games, the game is played on one team's home field. If there was no home field advantage, then what would be the probability that either team would win? If you said 50/50, that would be correct: Any team—road or home—would have an equal chance of winning. Applying this to the 92 games, this would mean that 46 times the road team won and 46 times the home team won, or at least very close to this outcome. If we let "W" represent the event "Win Championship Game," then 0.50 would represent P(W), or the probability of winning the championship game. If we let "H" represent "Playing Championship at Home," then if winning is independent of playing the game at home—that is, no home field advantage—the conditional probability of P(W|H) would also be 0.50, or just the probability of winning. Playing at home would offer no advantage. However, in reviewing these 92 championship games, 61 times the home team has won. This translates into roughly 2/3 of the time the home team has won. In other words, the conditional probability of winning the game given playing at home is 0.66, which is not equal to the 0.50 probability of winning if no home field advantage. From this, one could conclude that playing at home does offer a better chance of winning; a home field advantage exists in NFL championship games.

Expressions and Formulas

1. P(A) stands for the "probability that some event called A occurs."
2. P(A and B) or $P(A \cap B)$ is the probability of the "intersection of event A and event B.". This is where the two events overlap or the outcomes that occur in both event A and event B.
3. P(A or B) or $P(A \cup B)$ is the probability of the "union of event A and event B."
4. P(A') or $P(A^c)$ represents the probability of the "complement" of event A. This would be all outcomes that do NOT occur in event A.
5. P(A|B) represents a "conditional" probability. The probability event A occurs *given* event B has already occurred. Similarly, P(B|A) would be the conditional probability that event B occurs *given* event A has already occurred.
6. $P(A) = 1 - P(A^c)$, known as the complement rule.
7. P(A)*P(B) is the independence rule and can *only* be used if known that event A and event B are independent.
8. P(A or B) = P(A) + P(B) - P(A and B) is the addition rule for the union of event A and event B.
9. $P(A \mid B) = \frac{P(A \text{ and } B)}{P(B)}$ and $P(B \mid A) = \frac{P(A \text{ and } B)}{P(A)}$ are the formulas for conditional probabilities for event A and event B.
10. To show if two events are independent, we need to verify that any *one* of the following is true:
    1. P(A and B) = P(A)*P(B)
    2. P(A|B) = P(A)
    3. P(B|A) = P(B).

Probability Distributions 4

A **random variable** is a numerical characteristic of each event in a sample space, or equivalently, each subject in a population. Examples include:

Number of points scored in a NFL game;

Time it takes to complete a MLB game;

Heights of NBA players;

Goals scored in soccer.

These random variables are classified into two types: **discrete** or **continuous**. A **discrete random variable** has a countable set of distinct possible values, while a **continuous random variable** is such that any value (to any number of decimal places) within some interval is a possible value. A more defining difference would be that discrete random variables are counted and continuous random variables are measured. For instance, from the above examples, points and goals scored are counted, making them discrete; time and heights are measured, making them continuous.

What, then, is a **probability distribution**? Consider, for instance, how the heights of NBA players might be distributed. Do you think these would follow a uniform shape, meaning the height of each player is equally likely? Or possibly the heights mimic somewhat of a bell shape, where most heights fall within a certain range, then branch out in both directions at an equal pace. Or if considering total goals scored, does each outcome share an equal chance, or do you feel lower-scoring games are more likely to occur than higher-scoring games? A **probability distribution** of a random variable is used to illustrate or represent the probabilities of outcomes for a random variable.

4.1 Probability Distributions: Discrete Random Variable

Through the 2011 season, Aaron Rodgers has started 62 games as quarterback, all with the Green Bay Packers. If we define "X" to represent "number of touchdowns thrown in a game," we could set up a probability distribution table for this variable. (Remember, the choice of lettering is not important. We could just as easily use "Y" or "T" instead of "X.") Once we determine the total touchdowns thrown for each game, we can begin to form a table that depicts these results. We can take the number of games with a certain number of TD passes and divide this by 62, Rodgers's total games started, to get a per game TD probability. This represents the probability that a randomly selected game from his career would end with that total number of touchdowns thrown. Table 4.1 provides the total TDs thrown in a game and probability for each. This table is referred to as a **probability distribution table** (data found at www.nfl.com).

X = total TDs	0	1	2	3	4	5
P(X = x)	0.08	0.24	0.29	0.27	0.10	0.02

TABLE 4.1: Probability distribution table of touchdowns thrown in a game by Aaron Rodgers.

Interpreting the Table

A probability distribution table typically consists of two rows: the top row to represent the possible outcomes for some discrete random variable, and the bottom row to represent the probability of each of these outcomes occurring. In Table 4.1, the top row, X = total TDs is interpreted as "X is the total touchdowns Rodgers has thrown in a game." The range of this variable is from 0 to 5—hypothetically, he could have thrown for more than 5 but has yet to do so.

The second row in Table 4.1, P(X = x) gives the probability for each of these outcomes of 0 to 5. The capital "X" is used to represent the variable and the lower case "x" is used to represent an outcome of "X." For instance, from the table we could say, "The probability is 0.08 of a randomly selected game ending with him throwing zero touchdowns," or more concisely, we could write P(X = 0) = 0.08 where P(X = 0) is interpreted as "the probability that X is 0."

Remember that the probabilities come from how many of these 62 games ended with this total. For instance, the probability of 0.08 for X of 0 comes from 5 of the 62 games Aaron Rodgers started, resulting in him throwing zero touchdowns.

Questions (Answers after Last Question):

Q1. What total touchdown passes in a game were most likely to occur?

Q2. Are the total touchdown passes in a game mutually exclusive?

Q3. Are the number of touchdowns thrown in a game independent?

Q4. What do the entire probabilities sum to?

Q5. How would we find the probability if one of the X outcomes was not provided?

Q6. What is the probability that for a randomly selected game, the total touchdowns thrown would be 4 or more?

Q7. Looking at this distribution of touchdowns thrown, about what average number of TD passes would you expect per game for Aaron Rodgers?

Answers

A1. A total of 2, as this had the highest probability, i.e., outcome P(X = 2) was 0.29.

A2. Yes, for instance, for any one game, the total TD passes could not be 4 and 5 for the same game.

A3. No, they are not. Since they are mutually exclusive, then by rule they would be dependent. Remember, we are talking *within* a game and not across games!

A4. They sum to one. This is true for all probability distributions!

A5. We would add up the known probabilities and then subtract this sum from one.

A6. This is asking to find P(X > = 4) = P(X = 4 or X = 5) = P(X = 4) + P(X = 5) = 0.10 + 0.02 = 0.12. Conversely, we could use the complement rule and find this from 1 – P(X < 4) = 1 – P(X = 0 or X = 1 or X = 2 or X = 3) = 1 – (0.08 + 0.24 + 0.29 + 0.27) = 1 – 0.88 = 0.12.

A7. See below for **Expected Value**.

4.2 Expected Value (or Mean) of Discrete Random Variable

When you hear the term **expected value**, you should think mean. That is, expected value is another term for mean. And because we like to employ simplified notation, we use the expression E(X) to symbolize the "expected value of X." However, the calculation of this expected value for a discrete random variable does not follow the same math process as what you typically use for finding a mean. That is, when we are interested in the expected value for a discrete random variable, we do not simply add up the outcomes and divide by the total number of outcomes. We would not, for example, calculate the average per game touchdowns thrown by Rodgers simply by summing 0 through 5, then dividing by 6 (remember, the 0 counts!). This produces a result of 2.5, but assumes each outcome is equally likely. This is not the case, though, as some outcomes are more likely than others. Therefore, our calculations must account for this unequal weighting of outcomes.

As you can see from Table 4.1, games ending with 1, 2, or 3 touchdowns are much more likely than other outcomes. With this in mind, you might believe that the average, or mean, touchdowns thrown fall somewhere within this range, which would be less than the 2.5 mean calculated above—and you would be correct! So how does one account for these differing probabilities in calculating the expected value?

$$\textbf{Expected Value of X} = E(X) = \mu_X = \sum X_i P(X_i)$$

What is this expression? What this formula is doing is multiplying each outcome of X by its respective probability—this is the $X_iP(X_i)$—and then summing each of these. This is what is meant by the Σ.

Applying this formula to the data in Table 4.1, we find the expected number of touchdowns thrown by Aaron Rodgers in any given game to be:

$$E(X) = \Sigma X_iP(X_i) = (0)*(0.08) + (1)*(0.24) + (2)*(0.29) + (3)*(0.27) + (4)*(0.10) + (5)*(0.02)$$

$$= 0 + 0.24 + 0.58 + 0.81 + 0.40 + 0.10 = 2.13$$

As we stated earlier, we believed the mean should fall somewhere below our 2.5 calculated using the usual mean formula. However, when we look at this value of 2.13, we may realize that this outcome is not possible; a game cannot end with total touchdowns thrown that are not whole numbers. What should we do then when reporting this expected value when the result is not possible? Should we round the number to the nearest whole number? Round down? Round up? Just leave it alone? The correct choice is to leave the number "as is." We report the expected value as is, because this informs others that the mean is a little more than 2 but less than 3.

Naturally, you would not expect all games to end up with the same number; there has to be some variability. As we learned in Chapter 1, we can measure variability by calculating a variance and its square root partner, the standard deviation. Again, we have to modify our previously discussed formulas to account for outcomes not having equal weight (i.e., our outcomes do not have an equal probability of occurring).

NOTE: This average number of touchdowns per game started is the same as if we took Rodgers's total TD passes of 131 and divided by 62. So why not use this method? The reason is sometimes we are not provided that information. Instead, we might be given information such as: In Aaron Rodgers's career:

8% of the time he has thrown zero touchdowns in a game;

24% he has thrown one;

29% he has thrown two;

27% he has thrown three;

10% he has thrown four;

2% he has thrown five.

If presented in this manner, we would not know the total games or the total touchdown passes. We would need to create such a table as that provided in Table 4.1.

4.3 Variance and Standard Deviation of Discrete Random Variable

The formula for the variance of a discrete random variable, often referenced as Variance of X or Var(X) or V(X) or σ^2, can be found by the following:

$$\mathbf{Var(X) = \sigma^2 = \Sigma X_i^2 P(X_i) - [E(X)]^2}$$

As you can see, the formula uses some similar notation to that for finding the expected value. First, we multiply the squared outcomes by their respective probabilities and then sum—this is the $\Sigma X_i^2 P(X_i)$, then we subtract the square of the expected value. The standard deviation of X or SD(X) or S(X) or σ_x is simply the square root of the variance.

Applying this formula to the data in Table 4.1, we find the variance and standard deviation for the number of touchdowns thrown by Aaron Rodgers in a game to be:

$$= (0)^2 * (0.08) + (1)^2 * (0.24) + (2)^2 * (0.29) + (3)^2 * (0.27) + (4)^2 * (0.10) + (5)^2 * (0.02) - (2.13)^2$$

$$= 0 + 0.24 + 1.16 + 2.43 + 1.6 + 0.5 - 4.54 = 1.39$$

And the standard deviation of X = SD(X) = $\sigma_x = \sqrt{(\sigma^2)} = \sqrt{1.39} = 1.18$

Example

Table 4.2 provides the probability distribution for the total goals scored for each game of the 2010 men's World Cup.

X = total goals	0	1	2	3	4	5	6	7
P(X = x)	0.11	0.26	0.20	0.22	0.11	0.08	0.0	0.02

TABLE 4.2: Probability distribution table of goals scored in each game of 2010 men's World Cup.

Questions (Answers after Last Question):

Q1. What total goals were most likely to occur?

Q2. Are the total goals scored mutually exclusive?

Q3. Are the goals scored independent?

Q4. What do the entire probabilities sum to?

Q5. How would we find the probability if one of the X outcomes was not provided?

Q6. What is the probability that for a randomly selected game, the total goals scored were 5 or better?

Q7. What is the expected number of total goals per game?

Q8. What is the variance and standard deviation for goals scored?

Answers

A1. A total of 1, as this had the highest probability, i.e., outcome P(X = 1) was 0.26.

A2. Yes, for instance. for one game the total goals could not be 4 and 5 for the same game.

A3. No, they are not. Since they are mutually exclusive, then by rule they would be dependent. Consider the probability of scoring 2 goals in a game and the probability of scoring 3 goals in a game. If you knew that the game ended with 2 goals, what is the probability that the game ended with 3 goals? Since you know, i.e., given, that 3 goals were scored, then the probability of 2 goals being scored is 0. This P(2) = 0 does not equal P(2) = 0.20 and from probability rules, for these two events to be independent, then P(2|3) = P(2), and this is not the case!

A4. They sum to one. This is true for all probability distributions!

A5. We would add up the known probabilities and then subtract this sum from one.

A6. This is asking to find P(X > = 5) = P(5 or 6 or 7) = P(5) + P(6) + P(7) = 0.08 + 0.0 + 0.02 = 0.10 Conversely, we could use the complement rule and find this from 1 – P(X < 5) = 1 – P(0 or 1 or 2 or 3 or 4) = 1 – (0.11 + 0.26 + 0.22 + 0.20 +0.11) = 1 – 0.90 = 0.10.

A7. $E(X) = \sum X_i P(X_i) = (0)*(0.11) + (1)*(0.26) + (2)*(0.20) + (3)*(0.22) + (4)*(0.11) + (5)*(0.08) + (6)*(0.0) + (7)*(0.02) = 2.3$.

A8. $Var(X) = (0)^2*(0.11) + (1)^2*(0.26) + (2)^2*(0.20) + (3)^2*(0.22) + (4)^2*(0.11) + (5)^2*(0.08) + (6)*(0.0) + (7)*(0.02) - (2.3)^2 = 7.78 - 5.29 = 2.49$.

And the standard deviation of X or SD(X) = $\sqrt{Var(X)} = \sqrt{2.49} = 1.58$

4.4 Binomial Random Variable: A Special Discrete Random Variable

When a discrete random variable has only two possible outcomes, we call this a **binomial random variable**. The two outcomes are often referred to as "success" and "failure" outcomes. These two outcomes can exist "naturally," in that only two outcomes are possible, or the two outcomes can be created from a situation where there are more than two outcomes, but we label a particular subset of these outcomes as the success and the remainder as the failure. For instance, when someone shoots a foul shot, there are two possible (natural) outcomes: make or miss. One of these is called the success outcome and the other the failure, depending on which is the outcome of interest. On the other hand, consider the total-goals-scored scenario previously described. In that situation, there were eight possible outcomes. However, one might be interested in whether or not a game ended in a shutout, or X of 0. By doing this, we create a situation with only two outcomes—shutout or no shutout. The shutout would be the success and no shutout the failure.

When we have a binomial random variable, if certain conditions are satisfied, then we have a **binomial experiment**. When this occurs, we can take advantage and find probabilities for a host of possible outcomes. There are four conditions that must be satisfied in order for a binomial experiment to exist. These conditions are:

- There are a fixed number of trials (a fixed sample size).
- On each trial, the event of interest (success) either occurs or does not. That is, we have a binomial situation.
- The probability of success is the same on each trial.
- The trials are independent of one another.

Looking back at Table 4.2, what if our interest was simply whether or not any goals were scored (i.e., the game was a shutout)? This has two outcomes: zero goals or more than zero goals. If we consider a situation where we randomly select three games and want to find the probability that one of the three games was a shutout, can this scenario be considered a binomial experiment?

To answer this question, we need to verify that the four conditions have been satisfied.

- Is the number of trials fixed? Yes, we have a trial size of 3.
- In each trial, are there only two possible outcomes? Yes, either the game had no goals or there were goals scored.
- Is the probability of success the same for each trial? Yes, the probability of a shutout is 0.11 for each game.
- Are the trials independent? Yes, whether one game was scoreless would not affect whether the other games were scoreless.

With all four conditions satisfied, we have a binomial experiment.

4.5 Finding Probabilities in a Binomial Experiment

To understand finding probabilities for a binomial experiment, a good place to begin is to return to our original question of, "What is the probability that only one game of three was scoreless?"

If we consider the sample space, it would look like this, where S = shutout and N = no shutout: SNN, NSN, NNS as these are the only three possible outcomes where one of three games ended in a shutout. That is, we could have the first of the three games ending in a shutout, or the second game, or the third game. In each case, there is one shutout and two non-shutouts. Recall from Chapter 3 our discussion on independent events where we can multiply the probability of each outcome. This makes the probability of the first sample space, SNN, being found by P(S and N and N)) = P(S)* P(N)*P(N). From the table, we know that P(X = 0) is 0.11 and from the complement rule of Chapter 3, we know that P(X not equal to 0) = 1 – P(X = 0) or P(N) = 1 – P(S) = 1 – 0.11 = 0.89. Therefore, P(SSN) = 0.11*0.89*0.89 = 0.087. Applying this to all three possible scenarios in which one of three games ends in a shutout, we have:

P(SNN) = 0.11*0.89*0.89 = 0.087

P(NSN) = 0.89*0.11*0.89 = 0.087

P(NNS) = 0.89*0.89*0.11 = 0.087

Finally, since these three outcomes are mutually exclusive—for instance we could not experience the sequence SNN and NSN for the same three games selected—we add the probabilities to get the final solution. The probability of getting one shutout from three randomly selected games is found by:

P(SNN) + P(NSN) + P(NNS) = 0.087 + 0.087 + 0.087 = 0.261

Is there an easier way to calculate this, especially if we had a larger fixed number of trials (imagine if we used 10 games instead of 3!)? Yes, there is a simpler method called the **binomial formula**. In reviewing the above probability calculation, you may notice a pattern. We notice there are three possible outcomes of interest, each having the same probability. A formula would have to be able to account for the probability of each event, plus the number of times the event could occur. From above, this would mean the formula would need to calculate the 0.087 and that three such outcomes were possible. If we let "x" be the number of outcomes of interest (in our example, that is one) and "p" the probability of success (in our example is 0.11), then:

$$P(X = x) = \frac{n!}{x!(n-x)!} p^x (1-p)^{n-x}$$

where "!" is the symbol for "factorial" and means to take the n*(n-1)*(n-2)*....*1. For example, 3! would be 3*2*1, or 5! is 5*4*3*2*1 and note that 0! equals 1. Applying this formula to our soccer goal example, we have:

$$P(X = 1) = \frac{3!}{1!(3-1)!} * 0.11^1 * (1 - 0.11)^{3-1} = \frac{3*2*1}{1*(2*1)} 0.11 * 0.89^2$$

$$= 3 * 0.11 * 0.89^2 = 0.261$$

Exactly what we arrived at above.

As with any random situation, we would expect there to be an average result and some measure of variability. That is, if you repeatedly selected three games and counted the number of shutouts for the three selected, you would get varying results; some of the three games might have no shutouts up to all three games ending in a shutout. The question becomes, then, what is the expected number of successes and variance (standard deviation) for a binomial experiment? The expected value, or mean, is straightforward: With three games and 0.11 probability of a shutout for any one game, you might believe the natural thing to do would be to simply take the probability of success times the number of trials. Well, you would be correct! The mean, or expected value, and variance of a binomial experiment are:

$$\text{Mean: } E(X) = n*p$$

$$\text{Variance: } Var(X) = n*p*(1 - p)$$

Applying this to our example, the mean and variance would be:

$E(X) = 3*0.11 = 0.33$

$Var(X) = 3*0.11*0.89 = 0.298$, so standard deviation is $\sqrt{0.298} = 0.546$

Don't Forget About Zero and the Equal Sign!

Two common errors students make when learning about discrete random variables is the outcome of zero and the importance of the equal sign when discussing situations involving "fewer than" or "more than." First, they forget to consider the possibility that in the trials of a binomial experiment that one can get zero successes! When you look at the example above for number of shutouts over three games, a frequent mistake is to consider the number of possible successes is from one to three; that is, that in three randomly selected games, there can be either one, two, or all three games ending in a shutout. However, that is incorrect, as another possibility is that all three games end up with at least one goal being scored. You forget about the sample space NNN—none of the games end in a shutout. So remember the outcome of zero successes.

Second, there is a difference between "less than" and "less than or equal to." Saying "Less than 2" is not the same as saying "Less than or equal to 2." The latter includes the outcome of 2, while the former would only include outcomes below 2.

Example

If you've ever seen Shaquille O'Neal shoot foul shots, you probably realize that his chances are about fifty-fifty that he makes it (for his career, Shaq shot 52.7% from the foul line, so giving him a 50/50 chance is not far off!). Using this as his probability of success in making a foul shot, answer the following questions:

Q1. If he shoots until he makes two, would this be considered a binomial experiment?

Q2. If we look at 10 foul shots, what is the probability that he makes 4?

Q3. If we consider 10 foul shots, find the mean and standard deviation.

Q4. If we consider 10 foul shots, what is the probability he makes 4 or fewer?

Q5. If we consider 10 fouls shots, what is the probability he makes more than 4?

Answers

A1. No, this would not be binomial, as it violates Condition 1: There is not a fixed number of trials. Who knows how many attempts it will take until Shaq makes two!

A2. This is binomial, as we set the number of trials at 10 shots, the probability is 0.5 for each attempt, the attempts are independent, and there are two distinct outcomes per trial: miss or make. Applying our binomial formula for n = 10 and X = 4:

$$P(X = 4) = \frac{10!}{4!(10-4)!} * 0.5^4 * (1 - 0.5)^{10-4} = (210) * 0.0625 * 0.01563 = 0.20514$$

Or about a 20.5% chance that he makes exactly 4 of 10 foul shots.

A3. Since this is a binomial situation, the expected value and standard deviation are found by:

Mean: E(X) = n*p= 10*0.50 = 5

Variance: Var(X) = n*p*(1 – p) = 10*0.50*0.50 = 2.5 so standard deviation is

$$\sqrt{2.5} = 1.58$$

A4. To answer, we need to find $P(X \leq 4)$, as we are asked to find the probability he makes 4 or fewer. Note the difference between this notation and $P(X < 4)$, which would be interpreted as him making fewer than 4. Thus, we need to find the probability he makes 0, 1, 2, 3, or 4 shots.

$$P(X = 0) = \frac{10!}{0!(10-0)!} * 0.5^0 * (1 - 0.5)^{10-0} = (1) * 1 * 0.00098 = 0.00098$$

$$P(X = 1) = \frac{10!}{1!(10-1)!} * 0.5^1 * (1 - 0.5)^{10-1} = (10) * 0.5 * 0.00195 = 0.00975$$

$$P(X = 2) = \frac{10!}{2!(10-2)!} * 0.5^2 * (1 - 0.5)^{10-2} = (45) * 0.25 * 0.00391 = 0.04399$$

$$P(X = 3) = \frac{10!}{3!(10-3)!} * 0.5^3 * (1 - 0.5)^{10-3} = (120) * 0.125 * 0.00781 = 0.11715$$

P(X = 4) = 0.20514

$P(X \leq 4) = P(X = 0) + P(X = 1) + P(X = 2) + P(X = 3) + P(X = 4)$

= 0.00098 + 0.00975 + 0.04399 + 0.11715 + 0.20514

= 0.37701

A5. The question asks to find $P(X > 4)$ or $P(X \geq 5)$. One may answer hurriedly by calculating the probabilities for Shaq making 5, 6, 7, 8, 9, and 10 shots, which is a lot of work. However, if you stop and think for a second you should recognize that $P(X > 4)$ and $P(X \leq 4)$ are complement events. Therefore, $P(X > 4)$ plus $P(X \leq 4)$ equals one, and we can find $P(X > 4)$ by simply subtracting from one the probability that he makes 4 or fewer—which we found in the previous answer.

$P(X > 4) = 1 - P(X \leq 4) = 1 - 0.37701 = 0.62299$

4.6 Probability Distributions: Continuous Random Variable

A **continuous** random variable has possible values that form an interval—think of variables that are *measured* (e.g., height, weight). The probability distributions of continuous random variables are represented by a curve. The curve serves to specify the probability that the random variable falls in any particular interval of outcomes along this curve, where:

1. Any interval along the curve, i.e., the area under the curve for the interval, has a 0 to 1 probability.
2. The total probability, i.e., the total area under the curve, is equal to 1.

As we learned in Chapter 1, data can take on many shapes. For our discussion here, we will restrict our conversation to a specific bell-shaped distribution called the **normal distribution**. The normal distribution has a particular bell shape defined by two parameters: the mean and the standard deviation. These parameters are represented by the Greek letters μ and σ, respectively.[1] The normal distribution is arguably the most important statistical distribution, as many variables follow an approximately normal distribution. Since the normal distribution is bell shaped, the properties of the Empirical Rule apply. Therefore, the *probability* of observations falling with particular standard deviations from the mean are:

0.68 probability within one standard deviation of the mean: $\mu \pm \sigma$

0.95 probability within one standard deviation of the mean: $\mu \pm 2\sigma$

0.997 probability within one standard deviation of the mean: $\mu \pm 3\sigma$

4.7 Finding Probabilities for Normal Distribution

Recall from Chapter 1 where we introduced the **z-score**, defining this as the number of standard deviations an observation falls from the mean. The formula for finding z is:

$$z = \frac{\textit{observed score - mean}}{\textit{standard deviation}}$$

Using notation from this chapter, where X represents some observed score, we have:

$$z = \frac{X - \mu}{\sigma}$$

1 A somewhat complicated formula exists for the normal distribution curve, but its complexity is beyond this text. In a later chapter, we will address another bell-shaped distribution called the t-distribution, which takes on similar characteristics of the normal distribution when sample sizes are large.

The Z-Table in the appendix allows us to find probabilities for normally distributed data. The table provides **cumulative probabilities**—probabilities falling below some z-score. This table is referred to as a **Standard Normal Table**. By converting values that have a normal distribution to z-scores, the z-scores have a standard normal distribution with a mean of 0 and standard deviation of 1. This allows us to use the table for any data that has a normal distribution, even when the data is of different units (e.g., heights in inches and weight in pounds). If we convert these cumulative probabilities to a percentage by multiplying by 100%, we then have the **percentile** for this observed X score. In reading the table, we combine the leftmost column with the uppermost row (these are labeled Z). This combination produces z-scores that range from negative 3.09 to positive 3.09. Where the top rows and left columns meet within the table provides the cumulative probability for that two-decimal z-score. For example, if we combine the row -2.8 with the column .05, we find the probability 0.0022, or the cumulative probability is 0.0022 for a z-score of -2.85. The table is two pages: the first page for negative z-scores and the second page for positive z-scores.

Think for a second: You know the normal curve is a bell shape, indicating the mean is in the middle.

1. What do you think the z-score would be for an observed value at the median?
2. What would be the cumulative probability for this z-score?

If you consider the definition of a z-score, the number of standard deviations an observation is from the mean, then observing the mean would produce a z-score of zero. Next, since the mean is at the center of the normal distribution, this would imply that half of the area under the curve for the mean falls below it: or the probability is 0.5 for a z-score of zero. Looking at The Z-Table for a z-score of 0.00, we find exactly that: a cumulative probability of 0.5000![2]

When discussing probabilities there are four possible scenarios:

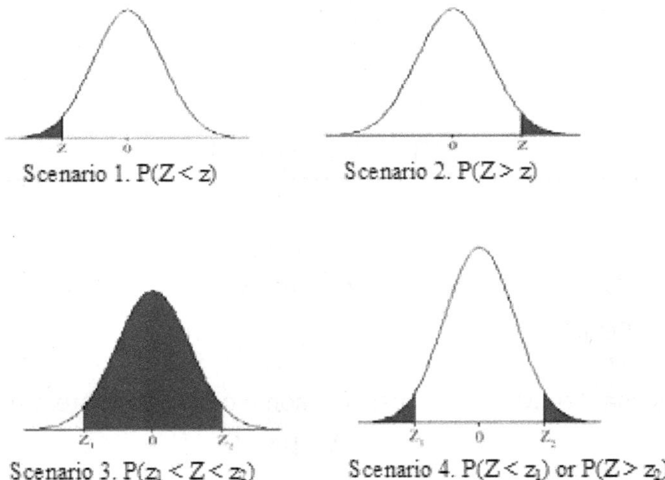

Scenario 1. $P(Z < z)$

Scenario 2. $P(Z > z)$

Scenario 3. $P(z_1 < Z < z_2)$

Scenario 4. $P(Z < z_1)$ or $P(Z > z_2)$

2 Unlike with discrete variables, the "equal" sign is not relevant with continuous variables. With the probability being represented by the area underneath a curve for an interval, the equal sign depicts an exact point: i.e., at exactly that observed value. Since the area under an exact point on a curve is zero, including the equal sign does not provide a difference. That is, the probability less than X is the same as the probability of less than or equal to X when discussing probabilities for continuous random variables.

Using this information, we can solve for finding various probabilities by first drawing a "picture," finding the z-score, and finding the cumulative probability.

To find the probabilities for each scenario:

Scenario 1. The cumulative probability from the Z-Table is the answer.

Scenario 2. Since the total area under the curve is one and the Z-Table provides the area under the curve to the left of some z-score, the probability for greater than is found by simply taking one minus the cumulative probability.

Scenario 3. The probability for "between two" z-scores is found by taking the difference between the cumulative probabilities for the z-scores.

Scenario 4. The "or" probability is a combination of Scenarios 1 and 2 and adding the two probabilities.

4.8 Example Using the Z-Table to Find Probabilities

For the 2011 season, the length of times (in minutes) for Phillies games was approximately normal, with a mean of 165.7 minutes and standard deviation of 18.63 minutes. These times reflected only regular season games that went only nine innings (data from www.baseball-reference.com).

Q1: What is the probability if you went to a Phillies game in 2011 that the game lasted less than two and a half hours (150 minutes)?

Q2: What is the probability that a 2011 Phillies game lasted more than three hours (180 minutes)?

Q3: What is the probability that a 2011 Phillies game lasted more than two and a half hours (150 minutes) but less than three hours (180 minutes)?

Q4: What is the probability that a 2011 Phillies game lasted less than two and a half hours (150 minutes) or more than three hours (180 minutes)?

A1: The question asks to find $P(X < 150)$, which is Scenario 1. We begin by converting this X of 150 to a z-score by:

$$z = \frac{X - \mu}{\sigma} = \frac{150 - 165.7}{18.63} = -0.84$$

From the Z-Table, we find the cumulative probability for z-score of negative 0.84 is 0.2005, or there is roughly a 20% chance that a Phillies game in 2011 ended in less than 150 minutes. Figure 4.1 provides a portion of the standard normal table with the cumulative probability for $z = -0.84$ marked.

FIGURE 4.1: Portion of standard normal table or Z-Table for time of Example Q1.

Z-Table: Standard Normal Cumulative Probabilities

Cumulative probability (area to LEFT) of Negative Z-values

Z	0.00	0.01	0.02	0.03	0.04	0.05	0.06	0.07	0.08	0.09
-3.0	0.0013	0.0013	0.0013	0.0012	0.0012	0.0011	0.0011	0.0011	0.0010	0.0010
-2.9	0.0019	0.0018	0.0018	0.0017	0.0016	0.0016	0.0015	0.0015	0.0014	0.0014
-2.8	0.0026	0.0025	0.0024	0.0023	0.0023	0.0022	0.0021	0.0021	0.0020	0.0019
....	...	...	...	...	...	...	...	...	...	...
...	...	...	...	...	...	...	...	...	...	...
-0.8	0.2119	0.2090	0.2061	0.2033	0.2005	0.1977	0.1949	0.1922	0.1894	0.1867

A2: The question asks to find P(X > 180) which is scenario 2. Converting this X of 180 to a z-score:

$$Z = \frac{X - \mu}{\sigma} = \frac{180 - 165.7}{18.63} = 0.77$$

From the Z-Table, we find the cumulative probability for z-score of 0.77 (remember to go to the second page for positive z-scores) is 0.7794, but this is the *cumulative* or "less than" probability, and our interest is in "greater than." To finish, we need to subtract this cumulative probability from one. (NOTE: Don't subtract one from the cumulative probability or you'll end up with a negative probability!) The answer is 0.2206 probability, or about a 22% chance that a Phillies game in 2011 took at least three hours. Figure 4.2 provides a portion of the standard normal table with the cumulative probability for z = 0.77 marked. FIGURE 4.2: Portion of standard normal table for Example Q2.

Z-Table (continued): Standard Normal Cumulative Probabilities

Cumulative probability (area to LEFT) of Positive Z-values

Z	0.00	0.01	0.02	0.03	0.04	0.05	0.06	0.07	0.08	0.09
0.0	0.5000	0.5040	0.5080	0.5120	0.5160	0.5199	0.5239	0.5279	0.5319	0.5359
0.1	0.5398	0.5438	0.5478	0.5517	0.5557	0.5596	0.5636	0.5675	0.5714	0.5753
0.2	0.5793	0.5832	0.5871	0.5910	0.5948	0.5987	0.6026	0.6064	0.6103	0.6141
...	...	...	...	...	...	...	...	...	...	...
...	...	...	...	...	...	...	...	...	...	...
0.7	0.7580	0.7611	0.7642	0.7673	0.7704	0.7734	0.7764	0.7794	0.7823	0.7852

A3: The question asks to find P(150 < X > 180), which is Scenario 3. Converting each X to a z-score:

$$Z_1 = \frac{X - \mu}{\sigma} = \frac{150 - 165.7}{18.63} = -0.84$$

$$Z_2 = \frac{X - \mu}{\sigma} = \frac{180 - 165.7}{18.63} = 0.77$$

From the Z-Table, we find the cumulative probability for each z-score, then take the difference (again, remember that probabilities cannot be negative so use the absolute value of the difference!). For a z-score of negative 0.84, we have cumulative probability of 0.2005 and for 0.77, we have a 0.7794 cumulative probability. Taking the difference, we arrive at a final probability of 0.5789, or about a 58% chance that a Phillies game in 2011 took between two and half to three hours.

A4: The question asks to find P(X < 150) or P(X > 180), which is Scenario 4. Converting each X to a z-score:

$$z_1 = \frac{X - \mu}{\sigma} = \frac{150 - 165.7}{18.63} = -0.84$$

$$z_2 = \frac{X - \mu}{\sigma} = \frac{180 - 165.7}{18.63} = 0.77$$

From the Z-Table, we find the cumulative probability for z_1 and one minus the cumulative probability for z_2, then add these results. For z_1 of negative 0.84, we have cumulative probability of 0.2005 and for z_2 of 0.77, we have 0.2206. Upon adding, we arrive at a final probability of 0.4211, or about a 42% chance that a Phillies game in 2011 took less than two and a half hours or more than three hours.

Going in Reverse: Finding an Observed Value for Some Given Cumulative Probability

There are times our interest might be in finding what observed score produces a certain percentile (i.e., the cumulative probability times 100%). Such instances require us to find the z-score for that cumulative probability, then solve for X in our z-score equation. This produces:

$$X = z^* \sigma + \mu$$

Example Using the Z-Table to Find Observed Score

Q: Staying with the Phillies 2011 game times, the top 1% of games lasted how long?

A: First, recall that this "top 1%" is the same as the 99th percentile. Therefore, the 99th percentile relates to a 0.9900 cumulative probability. Next, we need to find what z-score results in a 0.9900 cumulative probability. Often we will not find an exact match of this cumulative probability. When this occurs, a common practice is to take the "closest one" to this cumulative probability. Returning to the Z-Table, we set our goal of finding the cumulative probability (not z-score!) closest to 0.9900 (above or below doesn't matter—we are looking for closest). From the Z-Table, we find the closest cumulative probability is 0.9901, which corresponds to a z-score of 2.33. Figure 4.3 provides a portion of the standard normal table with the cumulative probability of 0.9901marked. Plugging this information into the above equation we get:

$$X = z^* \sigma + \mu = 2.33^* 18.63 + 165.7 = 209.11$$

Concluding that the top 1% of 2011 Phillies games lasted at least 209.11 minutes or roughly three and a half hours! Stated otherwise, approximately 99% of the 2011 Phillies games finished within three and a half hours.

FIGURE 4.3: Portion of standard normal table for finding z-score for some cumulative probability.

Z-Table (continued): Standard Normal Cumulative Probabilities

Cumulative probability (area to LEFT) of Positive Z-values

z	0.00	0.01	0.02	0.03	0.04	0.05	0.06	0.07	0.08	0.09
0.0	0.5000	0.5040	0.5080	0.5120	0.5160	0.5199	0.5239	0.5279	0.5319	0.5359
0.1	0.5398	0.5438	0.5478	0.5517	0.5557	0.5596	0.5636	0.5675	0.5714	0.5753
0.2	0.5793	0.5832	0.5871	0.5910	0.5948	0.5987	0.6026	0.6064	0.6103	0.6141
...	...	...	...	...	...	...	...	...	...	...
...	...	...	...	...	...	...	...	...	...	...
2.3	0.9893	0.9896	0.9898	0.9901	0.9904	0.9906	0.9909	0.9911	0.9913	0.9916

Expressions and Formulas

1. E(X) which stands for the "expectation of X," interpreted as the mean for x.
2. $E(X) = \sum X_i P(X_i)$ The equation for finding the expected value of X for a discrete random variable. If each outcome in X is equally likely, then this expected value of X is the same as finding the mean of X in Chapter 1.
3. $Var(X) = \sigma^2 = \sum X_i^2 P(X_i) - [E(X)]^2$ This is the formula for finding the variance of X for a discrete random variable. The symbol σ is the Greek letter sigma. As in Chapter 1, the standard deviation would be found by taking the square root of the variance.
4. For a binomial experiment, the probability of getting "x" number of outcomes in "n" trials is where $P(X=x)=\frac{n!}{x!(n-x)}p^x(1-p)^{n-x}$ "!" stands for "factorial." Remember, both 0! and 1! equal 1.
5. The mean of a binomial experiment is: E(X) = n*p, where "p" is the probability of getting a successful outcome in a binomial trial.
6. The variance of a binomial experiment is: Var(X) = n*p*(1 - p).
7. The symbol μ stands for the mean of some population set of data and is referred to as the population mean.

8. The symbol σ stands for the standard deviation of some population set of data and is referred to as the population standard deviation.

9. For data that follows a normal distribution (a particular type of bell shape), a z-score equals:

$$z = \frac{X - \mu}{\sigma}$$

This z-score looks just like the one from Chapter 1, and it is: we just replaced "observed score" with "X." As in Chapter 1, z-score is defined as the "number of standard deviations an observation is from the mean."

10. If we know the z-score or know the percentile which will provide us with the z-score, we can find the observation, or X value, for this percentile or z-score by:

$$X = z^{*}\sigma + \mu$$

Sampling Distributions 5

In Chapter 4, we focused on the distribution of a single observation sampled from some population as this related to the normal distribution. Throughout Chapters 1 and 2, we mentioned the idea of estimating population proportions and means. For instance, in Chapter 2, we discussed the *Ipsos* survey in estimating the proportion of Americans 12 and older who participated in fantasy sports. This idea of *estimating* relates to an important process in statistics called **statistical inference**. With statistical inference, we take sample data and make predictions and decisions regarding the population. The remainder of this text will concentrate on the study of inferential methods. Two key terms to begin this chapter are **statistics** and **parameters**.

Statistics are numerical summaries of a sample. Examples include the sample proportion and the sample mean. Simply put, when we take a sample and calculate the proportion or mean of the sample, this proportion or mean is referred to as the *sample proportion* or *sample mean*, depending on which is calculated. As we previously learned, these statistics can vary from sample to sample, as the subjects within each sample can vary. For example, say you and two friends randomly selected 20 students from your local university. Do you believe that each of you would select the exact same 20 students? Most likely no. True, some students may show up in more than one sample, but the likelihood that each of you selected the exact same 20 students is extremely unlikely.

Parameters are numerical summaries of a population. These values are fixed, such as the proportion of *all* Ohio State students who support the hiring of Urban Meyer, or the mean salary of *all* NFL players. Parameter values are fixed within that specific population.

However, these parameter values are seldom known; we use the statistics to estimate the parameter. This makes the statistic a **point estimate** of a parameter. A sample mean is considered a point estimate of a population mean, and a sample proportion is considered a point estimate of a population proportion.

In Chapter 4 we introduced the Greek symbol μ that represented a population mean. What about the population proportion? For our purposes, we will use the letter "p" to define a population proportion.[1] We will also introduce two new symbols for the statistics of sample mean and sample proportion. The symbols are $\bar{X}$ (read as "X bar") and $\hat{p}$ (read as "p hat"), respectively. To summarize:

1. $\bar{X}$ is a sample mean used as a point estimate of a population mean, μ.
2. $\hat{p}$ is a sample proportion used as a point estimate of a population proportion, p.

---

1 As mentioned throughout this book, some notation is not consistent. The importance is that you understand for your particular situation *what* symbol(s) are being used. Although the use of μ for the sample mean is common, you may find, for example, some texts/instructors use the pi symbol (π) as notation for the population proportion.

With the understanding that statistics will vary from sample to sample, the implication is that there exists a distribution for these samples: a **sampling distribution**. The concept of a sampling distribution can be thought of this way: You conduct a study where repeated samples are taken. From each sample, a statistic is calculated—a sample mean or sample proportion. As stated, these statistics will vary across the samples, thus having some variability. If you were to graph these sample statistics, they will have a shape—or distribution—and therefore have a mean and standard deviation. A measure of this standard deviation of a sample statistic is called the **standard error**. The standard error represents the standard deviation of a sample statistic.

Summary:

Population Distribution: This is the probability distribution from which a sample is drawn. The parameter values, such as the proportion p or mean μ, are usually unknown. For example, if we had a list of all NFL player salaries, the mean salary would define μ.

Sampling Distribution: This is the probability distribution of the sample statistic, such as the sample proportion $\hat{p}$ or sample mean $\bar{X}$. These statistics are used to estimate their respective population parameters. These sampling distributions will have a mean and standard deviation, the latter referred to as the ***standard error***.

As we will illustrate next, if certain conditions are met, the sampling distribution of two important statistics—the sample proportion and the sample mean—will take on an approximately normal distribution. When this happens, we can use our understanding of the normal distribution to find the probability of our sample producing a particular sample proportion or sample mean.

5.1 Sampling Distribution of Sample Proportion

Previously, we mentioned that if specific conditions are satisfied, the sampling distribution approximates a normal distribution. These conditions for sample proportions are:

Rules for sample proportions:

1. *If $np \geq 15$ **and** $n(1-p) \geq 15$, then the sample proportion can be assumed to follow an approximately normal distribution.*[2]
2. *The sample size is not more than 10% of the population when random sampling is done **without** replacement.*

2 Some texts and statistical software will use 5 or 10 instead of 15. The critical value of 15 was selected, as this provides a more conservative, or cautious, approach. In situations where the rule is not satisfied, the correct method to use is called an exact binomial. This method is beyond the scope of this text; however, most statistical software packages include this method.

Generally, sampling is done without replacement; that is, once a subject is selected, it is not put back into the population where it can be re-selected.

Also stated previously, we can use a mean and standard deviation to describe the sampling distribution for a sample statistic. In the case of the sample proportion, this mean and standard error are defined as follows:

When a random sample of size n is taken from a population for which there is a proportion, p, outcomes of interest, the sampling distribution of the sample proportion will have a:

$$\text{Mean} = p \text{ and standard error} = \sqrt{\frac{p(1-p)}{n}}$$

Explanation through Example

Back in Chapter 1, we talked about 2010 NFL player salaries. For purposes of discussion, we divide the 1257 salaries into one of two categories by the salary median ($855,000). By our understanding of the median splitting a data set in half, then the population proportion, p, of NFL players making above the median salary in 2010 was 0.5 or 50%.

If we were to randomly select a sample of 100 players from this set of data, then the distribution of the sample proportion for those making above the median salary has an approximately normal distribution with the following mean and standard error: (note that this size of 100 is not more than 10% of the population size—Rule 2 for proportions):

$$\text{Mean} = p = 0.50$$

$$\text{Standard error} = \sqrt{\frac{p(1-p)}{n}} = \sqrt{\frac{0.5(1-0.5)}{100}} = 0.05$$

This is not to say that our sample proportion would come out to exactly 0.50—we could get sample proportions of 0.49, 0.51, or even 0.30—but we would expect the sample proportion to be close to 0.5. By applying our *rules for sample proportions*, we can assume the sample proportion approximates a normal distribution, since 100*0.5 = 50 and 100*(1 - 0.5) = 50 are both at least 15.

What we see in Figure 5.1 is the result of taking a random sample of 100 salaries, calculating the proportion of the sample where the salary exceeds the mean, and repeating this process 5000 times. This produces 5000 sample proportions. Superimposed is a normal distribution having a mean of 0.5 and 0.05 standard error. As you can see, the distribution of these sample proportions approximates this normal distribution—the actual mean is 0.502 for these 5000 sample proportions.

With the sample proportions approximating a normal distribution, we would expect the Empirical Rule to hold and virtually all of the sample proportions would fall within three standard errors of the mean, or from 0.35 to 0.65. As Figure 5.1 illustrates, this concept holds true.

Given the time and cost involved in conducting a sample study, why would anyone go about repeating this sampling process more than once, let alone 5000 times? You would be correct if you said practically no one. In reality, we would only take one sample of size *n*. However, by understanding the theory behind how statistics are distributed, we can apply these theoretical concepts to make judgments regarding a population.

FIGURE 5.1 Sampling Distribution of 5000 Sample Proportions of size n = 100 taken from the 2010 NFL Salary Data.

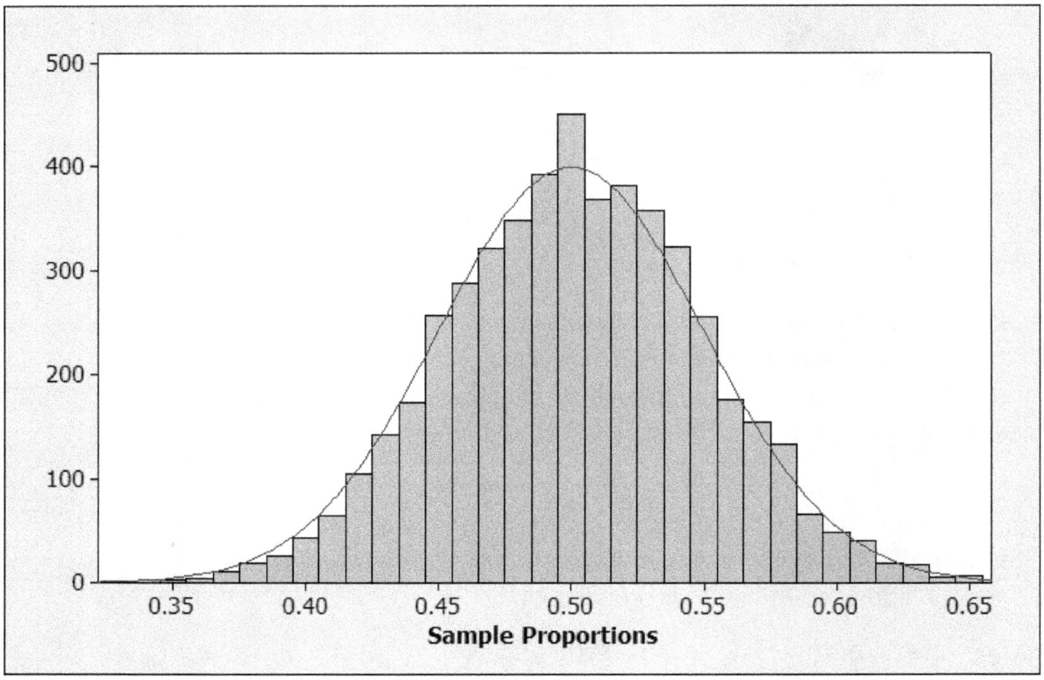

5.2 Application of Sample Distribution of Sample Proportion

In the prior section, we mentioned that although we would expect a mean proportion of 0.5, this is not meant to imply that our sample proportion would be 0.5; practically any sample proportion is possible. Applying what we learned about z-scores and The Z-Tables in Chapter 4, we can find the probability of attaining any sample proportion by converting the sample proportion to a z-score:

$$Z = \frac{\hat{p} - p}{\sqrt{\frac{p(1-p)}{n}}}$$ where $\hat{p}$ represents the sample proportion.

Example 1

Using the 2010 NFL player salaries broken down by median, what is the probability that if you took a random sample of size 100 from this data, you would get 40 players making above the median salary?

To start, recognize that the sample proportion is 0.4 found by 40/100. Converting this to a z-score, we get:

$$Z = \frac{\hat{p} - p}{\sqrt{\frac{p(1-p)}{n}}} = \frac{0.4 - 0.5}{\sqrt{\frac{0.5(1-0.5)}{100}}} = \frac{-0.1}{0.05} = -2.00$$

From the The Z-Table for a z-score of -2.0, we find the probability of 0.0228, or roughly a 2.3% chance of getting a sample proportion of 0.4 from a sample of 100. Figure 5.1 provides a portion of the standard normal table with the cumulative probability for z = -2.00 marked.

What if the sample size was 200 instead of 100? What would be the probability of getting a 0.4 sample proportion?

$$Z = \frac{\hat{p} - p}{\sqrt{\frac{p(1-p)}{n}}} = \frac{0.4 - 0.5}{\sqrt{\frac{0.5(1-0.5)}{200}}} = \frac{-0.1}{0.035} = -2.86$$

From the The Z-Table and locating a z-score of - 2.86, we find that the probability of getting a sample proportion of 40% or less for a sample size of 200 is 0.0021, or about a 0.2% chance. Figure 5.2 provides a portion of the standard normal table with the cumulative probability for z-scores -2.00 and -2.86 marked.

You can see the effect sample size has. As your sample size *increases*, the standard error will *decrease*.

FIGURE 5.2: Portion of standard normal table for Example 1.

Z-Table: Standard Normal Cumulative Probabilities

Cumulative probability (area to LEFT) of Negative Z-values

Z	0.00	0.01	0.02	0.03	0.04	0.05	0.06	0.07	0.08	0.09
-3.0	0.0013	0.0013	0.0013	0.0012	0.0012	0.0011	0.0011	0.0011	0.0010	0.0010
-2.9	0.0019	0.0018	0.0018	0.0017	0.0016	0.0016	0.0015	0.0015	0.0014	0.0014
-2.8	0.0026	0.0025	0.0024	0.0023	0.0023	0.0022	0.0021	0.0021	0.0020	0.0019
....	...	...	...	...	...	...	...	...	...	...
...	...	...	...	...	...	...	...	...	...	...
-2.0	0.0228	0.0222	0.0217	0.0212	0.0207	0.0202	0.0197	0.0192	0.0188	0.0183

Example 2

Back in Chapter 3, we referenced an *Ipsos* survey that stated 13% of U.S. adults ages 12 and older participated in fantasy sports during 2010. Assuming this proportion to be correct, if you took a sample of 1000 U.S. adults who were 12 or older in 2010 and asked them if they participated in fantasy sports during 2010, describe the shape of the distribution of the sample proportion as well as the mean and standard error.

First, we check the conditions for sample proportions. With a sample size of 1000 and population proportion of 13%, then np of 130 and n(1 - p) of 870 would exceed 15 and satisfy Rule 1. As to Rule

2, everyone should agree that the population of U.S. adults of at least age 12 is well over 10,000, which is ten times our sample size.

With the conditions satisfied, we can state that the distribution shape would be approximately normal and have the following mean and standard error:[3]

$$\text{Mean} = 0.13 \text{ and } S.E. = \sqrt{\frac{0.13(1-0.13)}{1000}} = 0.011$$

Example 3

For baseball players, having a 300 batting average (a proportion of hits per number of at-bats of 0.300) is often the goal. Achieving this point can result in increased salary, awards, etc. A regular player usually gets about 500 plate appearances per season.[4] In 2011, the mean batting average was 0.276 for those players who qualified (see www.mlb.com). Miguel Cabrera, of the Detroit Tigers, led the majors with a batting average of 0.344 based on 572 plate appearances. Describe the sampling distribution of Cabrera's average, including the shape, mean, and standard error. How unlikely was Cabrera to hit 0.344 based on the major league average?

The shape would be approximately normal, based on the rules for proportions, with a population proportion of 0.276 and sample size of 572. The mean and standard error would be:

$$\text{Mean} = 0.273 \text{ and } S.E. = \sqrt{\frac{0.276(1-0.276)}{572}} = 0.019$$

As to how unlikely Cabrera's average of 0.344 compared to the league average was, we can use the approximate normal methods (Z methods) to calculate the probability of a player achieving a 0.344 batting average.

$$Z = \frac{\hat{p}-p}{\sqrt{\frac{p(1-p)}{n}}} = \frac{0.344-0.276}{\sqrt{\frac{0.273(1-0.273)}{572}}} = \frac{0.068}{0.019} = 3.58$$

From the The Z-Table and locating a z-score of 3.58, Figure 5.3 provides a portion of the standard normal table, we notice that the table only goes to a z-score of 3.09, while the cumulative probability is 0.9990.

Since 3.58 is even further up the standard normal table, the cumulative probability for 3.58 would have to exceed 0.9990. This results in the probability of Cabrera hitting 0.344 being something less than 0.001, or less than a 0.1% chance.[5] Very unlikely indeed!

---

3 Remember, to keep the discussion clear between the measure of variability of a sample and sample statistic, we refer to the standard deviation of a sample statistic as the standard error (S.E.)

4 Currently, for a 162-game season, the minimum number of plate appearances is 502 in order to qualify for the batting title (see www.baseball-reference.com).

5 Using Minitab software to calculate the actual probability for this z-score, we achieve a probability of 0.00017 that Cabrera would hit 0.344.

FIGURE 5.3: Portion of standard normal table for Example 3.

Z-Table (continued): Standard Normal Cumulative Probabilities

Cumulative probability (area to LEFT) of Positive Z-values

Z	0.00	0.01	0.02	0.03	0.04	0.05	0.06	0.07	0.08	0.09
0.0	0.5000	0.5040	0.5080	0.5120	0.5160	0.5199	0.5239	0.5279	0.5319	0.5359
0.1	0.5398	0.5438	0.5478	0.5517	0.5557	0.5596	0.5636	0.5675	0.5714	0.5753
0.2	0.5793	0.5832	0.5871	0.5910	0.5948	0.5987	0.6026	0.6064	0.6103	0.6141
...	...	...	...	...	...	...	...	...	...	...
...	...	...	...	...	...	...	...	...	...	...
3.0	0.9987	0.9987	0.9987	0.9988	0.9988	0.9989	0.9989	0.9989	0.9990	0.9990

5.3 Sampling Distribution of Sample Mean

Just as we had conditions for sample proportions, we have conditions for sample means. In order for the distribution of a sample mean to be assumed to follow a normal or approximately normal distribution, ***only one*** of the following conditions must be satisfied.

Rules for sample means:

1. *If the population distribution is known to be normal or approximately normal, then the distribution of the sample mean, $\bar{X}$, is normal or approximately normal, regardless of sample size.*
2. *If the population distribution is skewed or unknown, the distribution of the sample mean, $\bar{X}$, is approximately normal, provided the sample size is sufficiently large.*

For Rule 2, "sufficiently large" is accepted to be a sample size of at least 30. This is referred to as the **Central Limit Theorem**.

When either rule is satisfied, the sample mean can be assumed to follow an approximately normal distribution with having a mean and standard error:[1]

Mean = population mean = μ and Standard Error = $\frac{\sigma}{\sqrt{n}}$, where σ is the population standard deviation. That is, $\bar{X}$ has a sampling distribution with a mean of μ and standard error $\frac{\sigma}{\sqrt{n}}$

1 There is a finite correction factor for this standard error when sampling is done without replacement. This is not covered in this text or many elementary statistics books. However, if the population size is at least 20 times the sample size, the finite correction factor can be ignored and the standard error as shown here can be applied.

Demonstration of Central Limit Theorem

As stated shown in Chapter 1, population distribution of 2010 NFL player salaries was heavily skewed to the right. The mean and standard deviation for this population was 1.7034 million dollars (or $1,703,400) and 2.112 million dollars (or $2,112,000). Figure 5.4 represents the distribution of these salaries. According to the Central Limit Theorem, if we take a sufficiently large sample ($N \geq 30$), the distribution of the sample mean will be approximately normal, with a mean equal to the population mean and a standard error equal to the population standard deviation divided by the square root of the sample size. Applying the Central Limit Theorem to the NFL salary data, if we took repeated samples of size 30 from this population and calculated the sample mean for each sample, then the distribution of these sample means should approximate a normal distribution with:

$$\text{Mean} = 1.7034 \text{ and Standard Error} = \frac{2.112}{\sqrt{30}} = 0.386$$

Figure 5.5 illustrates the result of this theorem for 2000 repeated samples of size 30 taken from the salary data. As you can see by comparing the two graphs, even for a heavily skewed population, the sample mean will approximate a normal distribution when the sample size is at least 30.

FIGURE 5. 4: Histogram of 2010 NFL player salaries in millions of dollars.

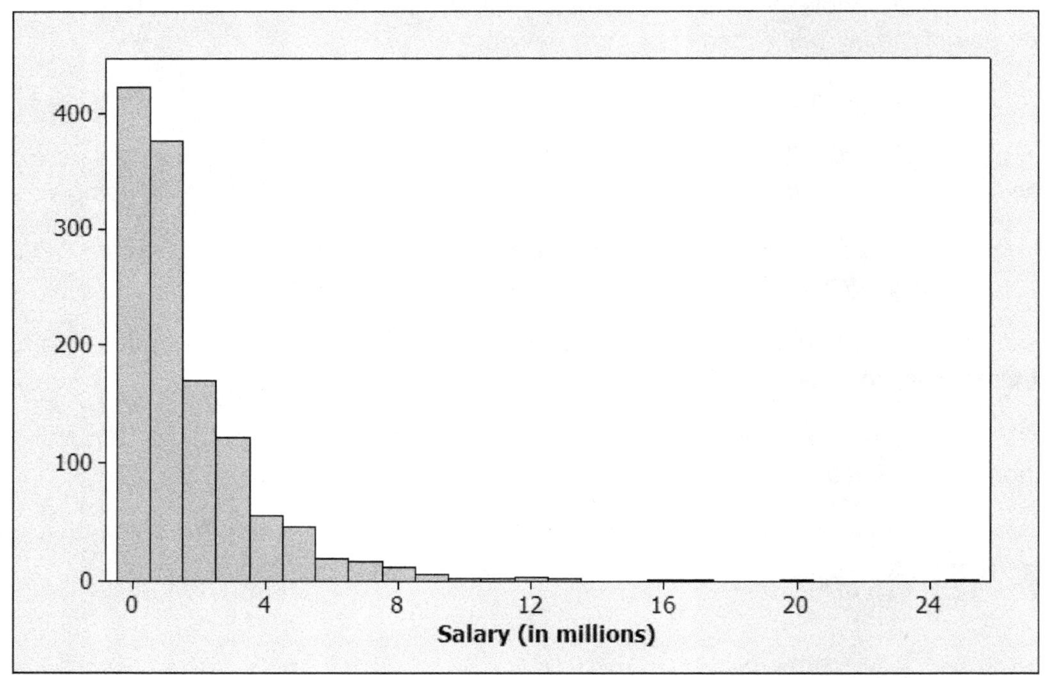

FIGURE 5.5: Histogram of 2000 sample means of size 30 randomly taken from the 2010 NFL player salary data. Superimposed is a normal distribution with mean of 1.7034 and standard deviation of 2.112.

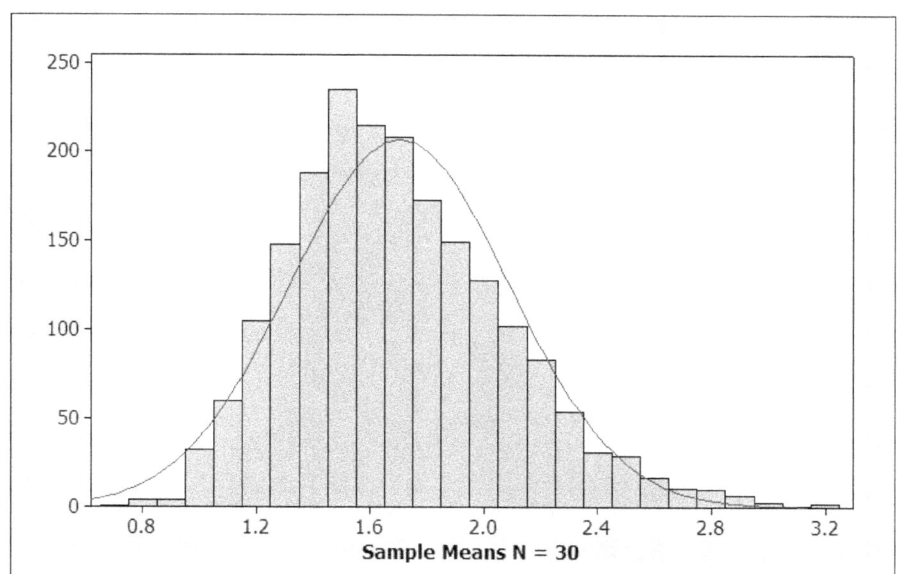

As we did with sample proportions, we can use these properties to find the distribution properties of a sample mean and to solve the probability of our sample data producing a particular sample mean.

5.4 Application of Sample Distribution of Sample Mean

When either of the sample mean rules are satisfied, we can apply Z methods to find probabilities of achieving a specified sample mean. The z-score for sample means is found by:

$$Z = \frac{\bar{X} - \mu}{\frac{\sigma}{\sqrt{n}}}$$

where $\bar{X}$ represents the sample mean, μ the population mean, and σ the population standard deviation.

Example 4

If a random sample of 50 players were selected from NFL salary data, describe the distribution, mean, and standard error for the sample mean.

Since we know the population data is heavily skewed, we employ Rule 2 for sample means, whereby having a sample size of at least 30 we can say that the distribution of the sample mean will be approximately normal. The distribution will take on the population mean of 1.7034 with a standard error of:

$$\text{S.E.} = \frac{\sigma}{\sqrt{n}} = \frac{2.112}{\sqrt{50}} = 0.30$$

Example 5

What is the probability that the sample mean for this sample of 50 is at least two million dollars? Converting the sample mean to a z-score:

$$Z = \frac{\bar{x} - \mu}{\frac{\sigma}{\sqrt{n}}} = \frac{2 - 1.7034}{\frac{2.112}{\sqrt{50}}} = \frac{0.2966}{0.30} = 0.99$$

From our The Z-Table, we find the cumulative probability for z of 0.99 is 0.8389. Since our interest was in "at least 2 million" we need the right tail probability for z found by 1 - 0.8389, resulting in a probability of 0.1611 or about a 16% chance that a sample of size 50 would produce a sample mean of two million dollars or more. Figure 5.6 provides a portion of the standard normal table with the cumulative probability for z = 0.99 marked.

FIGURE 5.6: Portion of standard normal table for Example 5.

Z-Table (continued): Standard Normal Cumulative Probabilities

Cumulative probability (area to LEFT) of Positive Z-values

Z	0.00	0.01	0.02	0.03	0.04	0.05	0.06	0.07	0.08	0.09
0.0	0.5000	0.5040	0.5080	0.5120	0.5160	0.5199	0.5239	0.5279	0.5319	0.5359
0.1	0.5398	0.5438	0.5478	0.5517	0.5557	0.5596	0.5636	0.5675	0.5714	0.5753
0.2	0.5793	0.5832	0.5871	0.5910	0.5948	0.5987	0.6026	0.6064	0.6103	0.6141
...	...	...	...	...	...	...	...	...	...	...
...	...	...	...	...	...	...	...	...	...	...
0.9	0.8159	0.8186	0.8212	0.8238	0.8264	0.8289	0.8315	0.8340	0.8365	0.8389

Example 6

From 2006 through 2011 (see www.pgatour.com), the average driving distance (in yards) for PGA players was approximately normal, with a mean of 281yards and standard deviation of 6.84 yards. Use this information to answer the following.

A. Find the z-score for a PGA golfer having an average driving distance of at least 283 yards.

B. Find the z-score for a random sample of 100 PGA golfers producing a sample mean of at least 283 yards.

C. Explain why an average distance of at least 283 yards for a single player would not be surprising, but a sample mean average of at least 283 yards for a sample of 100 players would be surprising.

Answers Example 6

A. To find the z-score, you need to recognize first that the data was assumed to follow an approximate normal distribution, and second, that in this situation we are only referencing *one* subject. Therefore, we use the z-score formula from Chapter 4:

$$Z = \frac{X - \mu}{\sigma} = \frac{283 - 281}{6.84} = \frac{2}{6.84} = 0.29$$

B. With the question referencing a sample mean of 100, we now employ the z-score formula from here in Chapter 5:

$$Z = \frac{\bar{X} - \mu}{\frac{\sigma}{\sqrt{n}}} = \frac{283 - 281}{\frac{6.84}{\sqrt{100}}} = \frac{2}{0.684} = 2.92$$

C. From our definition of **z-score**, where this explains *the number of standard deviations an observation is from the mean*, for a single player to average 283 yards would not be surprising, as this would translate to an observation being about 0.29 standard deviations above the mean. However, for the average driving distance of 100 players to produce a sample of mean of 283 yards would be unlikely, as this result corresponds to 2.92 standard deviations above the mean. We could also use the The Z-Table to find the probabilities for each z-score; we see that the probability of a single player averaging at least 283 yards is 0.3859 (from 1 - 0.6141), while the probability of a sample of 100 players averaging at least 283 yards is 0.0018 (from 1 - 0.9982). Figure 5.7 provides a portion of the standard normal table with the cumulative probability for z-scores 0.29 and 2.92 marked.

FIGURE 5.7: Portion of standard normal table for Example 6.

Z-Table (continued): Standard Normal Cumulative Probabilities

Cumulative probability (area to LEFT) of Positive Z-values

z	0.00	0.01	0.02	0.03	0.04	0.05	0.06	0.07	0.08	0.09
0.0	0.5000	0.5040	0.5080	0.5120	0.5160	0.5199	0.5239	0.5279	0.5319	0.5359
0.1	0.5398	0.5438	0.5478	0.5517	0.5557	0.5596	0.5636	0.5675	0.5714	0.5753
0.2	0.5793	0.5832	0.5871	0.5910	0.5948	0.5987	0.6026	0.6064	0.6103	0.6141
...	...	...	...	...	...	...	...	...	...	...
...	...	...	...	...	...	...	...	...	...	...
2.9	0.9981	0.9982	0.9982	0.9983	0.9984	0.9984	0.9985	0.9985	0.9986	0.9986

Expressions and Formulas

1. For some populations, the proportion of that population having a specific outcome of interest is "p."
2. The sampling distribution of the sample proportion, where n*p AND n*(1 - p) are at least 15, is approximately normal with the following mean and standard error:

   mean = p and standard error of $S.E. = \sqrt{\frac{p(1-p)}{n}}$
3. The sample proportion symbol is $\hat{p}$ and read as "p hat."
4. When the rules provided in No. 2 above are satisfied, then the z-score for a sample proportion, $\hat{p}$, is: $Z = \frac{\hat{p} - p}{\sqrt{\frac{p(1-p)}{n}}}$
5. Sample mean symbol is $\bar{X}$ and read as "X bar."
6. If a random sample is taken from a population data set that is normal or if the sample size is large enough (at least 30), then the sample distribution of the sample mean is approximately normal, with the following mean and standard error:

   mean = μ and standard error, $S.E. = \frac{\sigma}{\sqrt{n}}$, where μ is the population mean and σ is the population standard deviation.
7. If the sample mean can be assumed to follow a normal distribution based on No. 6, then the z-score for a sample mean, $\bar{X}$ is: $Z = \frac{\bar{X} - \mu}{\frac{\sigma}{\sqrt{n}}}$

Confidence Intervals 6

Two designs for producing data are sampling and experimentation, both of which should employ randomization. As we have already learned, one important aspect of randomization is to control for bias. Now we will see another positive. Because chance governs our selection (think of guessing whether a flip of a fair coin will produce a head or a tail), we can make use of probability laws—the scientific study of random behavior—to draw conclusions about an entire population from which the subjects originated. This is called **statistical inference**.

Back in Chapter 2, we defined population and sample. Then, in Chapter 5, we defined parameter, statistic, and point estimate. Due to their importance, we'll revisit these terms.

Parameter: A summary measure that describes the population. It is fixed, but we rarely know it. Examples include the proportion of American adults who participate in fantasy sports, proportion of NFL home teams who win, the mean height of all NBA players, or the mean attendance for major league baseball games.

Statistic: A summary number that describes the sample. This value is known since it is produced by our sample data, but can vary from sample to sample. For example, if we calculated the proportion of a sample of 1000 American adults who said they participated in fantasy sports last year, this would be the sample proportion. If we took another random sample of 1000 American adults asking the same question, the proportion who said yes for this sample would most likely vary from the proportion of our first sample, since the two samples themselves would consist of different American adults. This fluctuation in sample statistics is called **sampling error**.

Point Estimate: Certain statistics are referred to as point estimates of a parameter. For example, the sample proportion is labeled a point estimate of the population proportion; sample mean is called a point estimate of the population mean. In statistical notation we are saying:

$$\hat{p} \xrightarrow{\text{estimates}} p$$
$$\bar{x} \xrightarrow{\text{estimates}} \mu$$

Examples

1. A survey is carried out at a university to estimate the proportion of undergraduate students who attended that university's sporting events within the past year. One thousand undergraduate students from that campus are randomly selected and asked whether they attended an on-campus sporting event within the past year. The **population** is all of the undergraduates at that university campus. The **sample** is the group of 1000 undergraduate students selected. The **parameter** is the proportion of all undergraduate students at that university campus who attended an on-campus sporting event within the past year. The **statistic** is the proportion of the

1000 sampled undergraduates who said yes to having attended an on-campus sporting event within the past year.

2. A study is conducted at a university to estimate the mean number of on-campus sporting events that undergraduate students attended within the past year. Five hundred undergraduate students from that campus are randomly selected and asked how many on-campus sporting events they attended within the past year. The **population** is all of the undergraduates at that university campus. The **sample** is the group of 500 undergraduate students selected. The **parameter** is the mean number of on-campus sporting events attended by all undergraduate students at that university campus within the past year. The **statistic** is the mean number of on-campus sporting events attended by the 500 undergraduates sampled.

Ultimately, we will take these statistics and use them to draw conclusions about parameters. This is statistical inference. We begin our discussion with **confidence intervals**. Confidence intervals are primarily used to estimate some unknown parameter by providing a range of values (the interval) with some degree of confidence that the interval correctly contains the parameter. A confidence interval follows the format:

Sample Statistic ± *Margin of Error*

In Chapter 2, we discussed margin of error and provided a conservative margin of error for proportions using the equation $\frac{1}{\sqrt{n}}$. In this chapter, we will provide a more exact margin of error for a proportion, plus discuss confidence intervals for a mean. The **margin of error** provides a measure of accuracy of our point estimate in estimating a parameter. The margin of error will consist of two pieces: a multiplier and a standard error. The multiplier will be based on the distribution of the sample statistic, and the standard error will be for that sample statistic. The margin of error, therefore, has this setup:

Margin of Error = *Multiplier x Standard Error*

In Chapter 5, we learned the standard errors for the sample proportion and sample mean. At that time, we also learned about the distribution of these statistics when satisfying certain rules.

6.1 One-Proportion Confidence Intervals[1]

In Chapter 5, we pointed out that if $np \geq 15$ **and** $n(1\text{-}p) \geq 15$, the sample proportion can be assumed to follow an approximately normal distribution. The problem with this rule as it pertains to the current chapter is that we don't *know* the population parameter, p; we want to estimate it. Because of this, the rule changes by inserting the sample proportion, $\hat{p}$, for p. To simplify this, we recall another prior concept, the binomial random variable (see Chapter 4—yet another concept on which we build!).

1 In Chapter 8, we will look at intervals for comparing two populations. Starting from Chapter 6 onward, the key for students to be successful will be to understand which statistical method applies to a given situation.

From the binomial, we should recognize that $n\hat{p}$ is the number of "successes" and $n(1 - \hat{p})$is the number of "failures."

When this rule is satisfied, a valid confidence interval for a proportion, p, can use a multiplier from the standard normal (Z) distribution and standard error as follows:

$$S.E. = \sqrt{\frac{\hat{p}(1 - \hat{p})}{n}}$$

We just substituted the sample proportion for the population proportion in our standard error equation from Chapter 5.

The multiplier will come from the standard normal distribution. Some common levels of confidence are 90%, 95%, 98%, and 99%, with respective Z-multipliers of 1.65, 1.96, 2.33, and 2.58. To understand how we arrived at these multiplier values, we will examine the 1.96 used for a 95% confidence interval.

From the The Z-Table, we want to find the probability between a negative and positive z-score associated with our level of confidence; for 95% this is 0.95. As Figure 6.1 illustrates, we would find a 0.95 probability between negative and positive 1.96 z-scores. Similar applications would demonstrate the other multipliers. Table 6.1 organizes these multipliers for one-proportion confidence intervals.

FIGURE 6.1: Illustration for z-multiplier being 1.96 for 95% confidence interval.

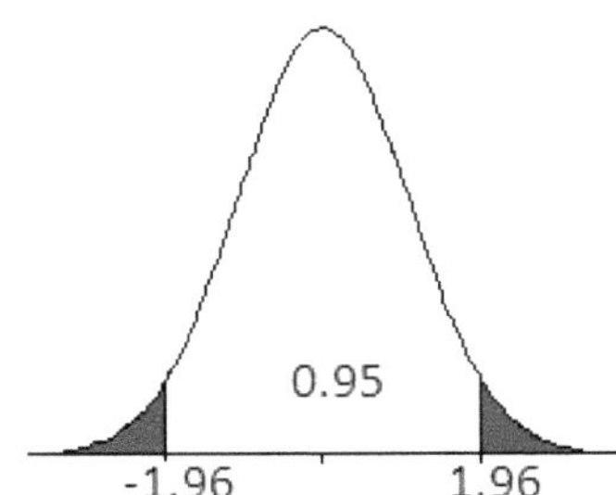

Confidence	Z Multiplier
90%	1.65
95%	1.96
98%	2.33
99%	2.58

TABLE 6.1: Z-multipliers for specific levels of confidence.

Putting these pieces together, our one-proportion confidence interval takes the form:

$$\hat{p} \pm Z* \sqrt{\frac{\hat{p}(1 - \hat{p})}{n}}$$

where $Z*\sqrt{\frac{\hat{p}(1-\hat{p})}{n}}$ is the margin of error for the confidence interval.

From these equations and Table 6.1, we should recognize a few general concepts about these intervals, as well as confidence intervals in general:

1. As the level of confidence increases, the margin of error increases, making our intervals wider.

2. As the sample size increases, the standard error decreases, making our intervals narrower.

Students will sometimes confuse *confidence* with *precision*. Remember that we are trying to estimate with some degree of confidence a range of values for which the parameter will fall. With a wider interval providing more possible outcomes, we in turn are more confident this wider interval is correct in capturing the parameter.

Interpreting the Confidence Interval

With the primary purpose of our confidence interval being the estimation of a population parameter, our interpretation should reflect this mission. When we say we have 95% confidence or 99% confidence, etc., the meaning relates to how our calculation methods would perform over a long-run series of samples. That is, if we repeated the interval calculation over and over and over, each time using another random sample, we would be confident that 95%, or 99%, or whatever percent of the intervals calculated, would contain the parameter. For instance, if we took a random sample of 1000 subjects from a population, calculated a 95% confidence interval, and repeated this process again and again—say 100 times, we would expect 95 of these 100 confidence intervals to contain the parameter value.

In general, we can build a template of interpreting confidence intervals. This template is:

*We are (*insert percent*) confident that (*insert parameter and population of interest*) is from (*insert interval bounds*).*

Example 1: Winning at Home in the NFL

What is the proportion of NFL teams that win at home? Construct 95% and 99% confidence intervals to estimate this proportion. Using the results of the 2011 NFL season as a sample of all NFL games (see www.repole.com), we find that the home team won 145 of 256 games played.

First, we check to verify that the z-methods can be applied in the calculation of the interval. With 145 "successes" and 111 "failures," both being at least 15, the methods can be used.

Next, we complete our confidence interval equation using 145/256 = 0.57 as our sample proportion and using the appropriate z-multiplier from Table 6.1.

95% Confidence Interval

$$\hat{p} \pm Z*\sqrt{\frac{\hat{p}(1-\hat{p})}{n}} = 0.57 \pm 1.96*\sqrt{\frac{0.57(1-0.57)}{256}} = 0.57 \pm 1.96*0.031 = 0.57 \pm 0.06$$

Adding and subtracting from 0.57 the margin of error of 0.06, we get a 95% confidence interval of (0.51 to 0.63) or from (51% to 63%). Table 6.2 gives Minitab output for this calculation.

99% Confidence Interval

Here all we need to change is the z-multiplier from 1.96 to 2.58, resulting in a 0.08 margin of error. Adding and subtracting from 0.57 this margin of error of 0.08, we get a 99% confidence interval of (0.49 to 0.65) or from (49% to 65%).

Notice that the 99% confidence interval is *wider* than the 95% confidence interval.

Finally, we need to put into context—give an interpretation of—our resulting confidence interval.

Interpretation: We are 95% confident that the proportion of NFL games won by the home team is from 0.51 to 0.63, or from 51% to 63%. At the 99% level of confidence, this range changes to 0.49 to 0.63, or 49% to 63%.

Variable	X	N	Sample p	95% CI
Home Won	145	256	0.566406	(0.505700, 0.627113)
	↑ Count	↑ Sample Size	↑ Sample Proportion	↑ 95% Confidence Interval Endpoints

TABLE 6.2: Annotated Minitab output for 95% confidence interval for Example 1—proportion of NFL teams winning at home.

Example 2: Winning at Home in College Football

What is the proportion of college football teams that win at home?[3] Construct a 90% confidence interval to estimate this proportion. Using the results of the 2011 college season (Division 1 or FBS) as a sample of all Division 1 college games (see www.repole.com), we find that the home team won 483 of 748 games played.

To start, we check to verify that the z-methods can be applied in the calculation of the interval. With 483 "successes" and 265 "failures," both being at least 15, the methods can be used.

Next, we complete our confidence interval equation, using 483/748 = 0.65 as our sample proportion and using the appropriate z-multiplier from Table 6.1.

2 Home games played at home team's campus stadium. For example, the Georgia-Florida game is played in Jacksonville, so this game was not included.

90% Confidence Interval

$$\hat{p} \pm Z^* \sqrt{\frac{\hat{p}(1-\hat{p})}{n}} = 0.65 \pm 1.65 * \sqrt{\frac{0.65(1-0.65)}{748}} = 0.65 \pm 1.65 * 0.017 = 0.65 \pm 0.03$$

Adding and subtracting from 0.65 the margin of error of 0.03, we get a 90% confidence interval of (0.62 to 0.68) or from (62% to 68%). Table 6.3 gives annotated Minitab output for this calculation.

Interpretation: We are 90% confident that the proportion of Division 1 college games won by the home team is from 0.62 to 0.68, or from 62% to 68%.

```
Variable     X     N    Sample p           90% CI
Home Won    483   748   0.645722    (0.616956,  0.674487)
             ↑     ↑       ↑                  ↑
          Count  Sample  Sample     90% Confidence Interval Endpoints
                  Size   Proportion
```

TABLE 6.3: Annotated Minitab output for 90% confidence interval for Example 2—proportion of college football teams winning at home.

Example 3: Beating the Point Spread in the NFL

What is the proportion of NFL home teams beating the point spread? Construct a 98% confidence interval to estimate this proportion. Using the results of the 2011 NFL season as a sample of all NFL games (see www.repole.com), we find that the home covered 123 of 256 games played.

Again, we check to verify that the z-methods can be applied in the calculation of the interval. With 123 "successes" and 133 "failures," both being at least 15, the methods can be used.

Next, we complete our confidence interval equation using 123/256 = 0.48 as our sample proportion and using the appropriate z-multiplier from Table 6.1.

98% Confidence Interval

$$\hat{p} \pm Z^* \sqrt{\frac{\hat{p}(1-\hat{p})}{n}} = 0.48 \pm 2.33 * \sqrt{\frac{0.48(1-0.48)}{256}} = 0.48 \pm 2.33 * 0.031 = 0.48 \pm 0.07$$

Adding and subtracting from 0.48 the margin of error of 0.07, we get a 98% confidence interval of (0.41 to 0.55) or from (41% to 55%). Table 6.4 gives annotated Minitab output for this calculation.

Interpretation: We are 98% confident that the proportion of NFL home teams that cover the spread is from 0.41 to 0.55, or from 41% to 55%.

```
Variable      X    N    Sample p          98% CI
Covered     123  256    0.480469   (0.407826, 0.553112)
             ↑    ↑        ↑                ↑
          Count Sample   Sample    98% Confidence Interval Endpoints
                 Size   Proportion
```

TABLE 6.4: Annotated Minitab output for 98% confidence interval for Example 3—proportion of NFL home teams that beat the point spread.

6.2 Finding Sample Size for Estimating a Population Proportion

When someone begins a study to estimate a population parameter, they typically have an idea of how confident they want to be in their results and within what degree of accuracy—the margin of error. They get started with a set level of confidence and a specified margin of error. We can use these pieces to determine a minimum sample size needed to produce these results by using algebra to solve for n in our margin of error:

$$n = \frac{z^2\hat{p}(1-\hat{p})}{M.E.^2}$$

Conservative estimate: If we have no preconceived idea of the sample proportion (e.g., previous study results), then a conservative estimate (that is, guaranteeing the largest sample size calculation) is to use 0.5 for the sample proportion. For example, if we wanted to calculate a 95% confidence interval with a margin of error equal to 0.04, then a conservative sample size estimate would be:

$$n = \frac{z^2\hat{p}(1-\hat{p})}{M.E.^2} = \frac{1.96^2(0.5)(1-0.5)}{0.04^2} = 600.25$$

Since this is the *minimum* sample size and we cannot get 0.25 of a subject, we **round up**. This results in a sample size of 601.

Estimate when proportion value is hypothesized: If we have an idea of a proportion value, then we simply plug that value into the equation. Note that using 0.5 will always produce the largest sample size, and this is why it is called a conservative estimate.

6.3 One Mean Confidence Intervals

Previously, we considered confidence intervals for 1-proportion and our multiplier in our interval used a z-value. But what if our variable of interest is a quantitative variable—points scored, margin of victory, time to complete baseball game—and we want to estimate the population mean? In such a situation, proportion confidence intervals are not appropriate, since our interest is in a **mean** amount and not a proportion.

The solution involves similar techniques for a proportion confidence interval, except we will be interested in estimating the population mean, μ, by using the sample mean, $\bar{X}$. A more significant change comes in choosing our multiplier.

In Chapter 5, we stated that under certain conditions—population is normally distributed or if sample size is at least 30—the sample mean will approximate a normal distribution. One underlying factor for this application was in knowing the population standard deviation, **σ**. In practice, knowing the population standard deviation is unlikely. As a result, this, too, is estimated, and we use the sample standard deviation, S or SD. This correction has an effect on the distribution of the sample mean, as this introduces extra error, especially when samples are small. To offset this increased error, we use **t-scores** in place of z-scores. These t-scores come from a t-distribution, which is similar to the standard normal distribution from which we get the z-scores. The similarities are that the t-distribution is symmetrical and centered on 0. A difference is that the standard deviation of the t-distribution is somewhat larger than 1; the standard deviation of the standard normal (Z) distribution. The specific standard deviation value depends on what is called the **degrees of freedom** (**df**). The df will be based on the sample size and are found by **df = n − 1**.

Following our general format for a confidence interval: *Sample Statistic* ± *Margin of Error* and inserting our notation for sample means and using t-scores for multipliers, our confidence interval for one mean will take on the following:

$$\bar{x} \pm t^* \frac{s}{\sqrt{n}}$$

where $t^* \frac{s}{\sqrt{n}}$ is the margin of error for a one-mean confidence interval and $\frac{s}{\sqrt{n}}$ is the standard error where we replaced the population standard deviation, σ, with the sample standard deviation, *s*.

Using a T-Table to Find Multipliers for Confidence Intervals about a Mean[3]

Below in Figure 6.2 is an portion from the T-table in the appendix. To read this table to find multipliers, we combine the correct degree of freedom (df) row corresponding to the confidence level of interest. For example, if the sample size is 8 and we are interested in a 95% confidence interval for a mean, the correct t-value multiplier would come from matching the df for 7 (from 8 - 1) with the column under 95% confidence level. The corresponding t-value of 2.365 would serve as the multiplier:

3 The df in a T-table typical run from 1 to 30 then move up in various increments. This is done as the change in t-values is minimal after reaching sample size of 30. A common practice for df not on the table is to use the closer one without exceeding. As the table indicates, the last row for infinity, ∞, has t-scores that match the Z-table. As sample size increases, the t-distribution approaches the standard normal distribution.

FIGURE 6.2: Portion of T-table.

T-Table: t Distribution Confidence Interval and Critical Values

	Confidence Level					
	80%	90%	95%	98%	99%	99.8%
	Right Tail Probability					
df	$t_{0.10}$	$t_{0.05}$	$t_{0.025}$	$t_{0.01}$	$t_{0.005}$	$t_{0.001}$
1	3.078	6.314	12.706	31.821	63.657	318.289
2	1.886	2.920	4.303	6.965	9.925	22.328
..	...	...	...	...	...	...
..	...	...	...	...	...	...
7	1.415	1.895	2.365	2.998	3.499	4.785

Example 4: Average Points Scored in an NFL Game

What is the mean number of points scored in an NFL game? Construct 95% and 99% confidence intervals to estimate this proportion. Using the results from a random sample of 25 games played in 2011 as a sample of all NFL games, the sample produced a mean of 50.12 points and standard deviation of 11.71 points.

First, we check to verify that the t-methods can be applied in the calculation of the interval. With a sample size of 25, we would need to assume that the population of points scored in an NFL game were approximately normal, since our sample size is not large enough (at least 30) to apply the Central Limit Theorem.

Next, we complete our confidence interval equation using 50.12 as the sample mean and 11.71 as the sample standard deviation. From the T-table, a portion of which is provided in Figure 6.3, using degrees of freedom (df) of 24, we find the t-multipliers for 95% confidence and 99% confidence are 2.064 and 2.797, respectively.

95% Confidence Interval

$$\bar{x} \pm t^* \frac{s}{\sqrt{n}} = 50.12 \pm 2.064 * \frac{11.71}{\sqrt{25}} = 50.12 \pm 2.064 * 2.342 = 50.12 \pm 4.83$$

Adding and subtracting from 50.12 the margin of error of 4.83, we get a 95% confidence interval of (45.29 to 54.95).

FIGURE 6.3: Portion of T-Table for 95% and 99% confidence intervals in Example 4.

T-Table: t Distribution Confidence Interval and Critical Values

	Confidence Level					
	80%	90%	95%	98%	99%	99.8%
	Right Tail Probability					
df	$t_{0.10}$	$t_{0.05}$	$t_{0.025}$	$t_{0.01}$	$t_{0.005}$	$t_{0.001}$
1	3.078	6.314	12.706	31.821	63.657	318.289
2	1.886	2.920	4.303	6.965	9.925	22.328
..	...	...	...	...	...	...
..	...	...	...	...	...	...
24	1.318	1.711	2.064	2.492	2.797	3.467

99% Confidence Interval

Here all we need to change is the t-multiplier from 2.064 to 2.797, resulting in a 6.55 margin of error. Adding and subtracting from 50.12 this margin of error of 6.55, we get a 99% confidence interval of (43.57 to 56.67). Table 6.5 provides annotated Minitab output for the 95% and 99% confidence intervals.

Notice that the 99% confidence interval is *wider* than the 95% confidence interval.

Finally, we need to put into context—give an interpretation of—our resulting confidence interval.

Interpretation: We are 95% confident that the mean number of points scored in NFL games is from 45.29 to 54.95 points. At the 99% level of confidence, this estimation changes to 43.57 to 56.67 points.

```
Variable         N    Mean   StDev  SE Mean       95% CI            99% CI
Total Points    25   50.12   11.71    2.34   (45.29, 54.95)  (43.57, 56.67)
                 ↑     ↑       ↑        ↑           ↑                ↑
            Sample  Sample  Standard  Standard  Endpoints for 95%  Endpoints for 99%
             Size    Mean   Deviation  Error   Confidence Interval Confidence Interval
```

TABLE 6.5: Annotated Minitab output for 95% and 99% confidence interval for Example 4—mean points scored in an NFL game.

Example 5: Average Margin of Victory for the Home Team in the NFL

What is the mean margin of victory for the home team in the NFL? Construct a 90% confidence interval to estimate this mean. Using the results of the 2011 NFL season as a sample of all NFL games, we found that the home team won 145 times, with a mean margin of victory of 13.5 points and standard deviation of 10.7 points (see www.nlf.com).

We begin by verifying that the t-methods can be applied in the calculation of the interval. With a sample size of 145, we can assume the sample mean will follow an approximately normal distribution based on the Central Limit Theorem. The t-methods would apply.

Next, we complete our confidence interval equation, using 13.5 as the sample mean and 10.7 as the sample standard deviation. From the T-table, a portion of which is found in Figure 6.4, the degrees of freedom (df) are 144. However, since this df value is not listed, we select the closest without exceeding, or df of 100. This gives us a t-multiplier of 1.660 for a 90% confidence interval.

FIGURE 6.4: Portion of T-Table for 90% confidence interval in Example 5.

T-Table: t Distribution Confidence Interval and Critical Values

	Confidence Level					
	80%	90%	95%	98%	99%	99.8%
	Right Tail Probability					
df	$t_{0.10}$	$t_{0.05}$	$t_{0.025}$	$t_{0.01}$	$t_{0.005}$	$t_{0.001}$
1	3.078	6.314	12.706	31.821	63.657	318.289
2	1.886	2.920	4.303	6.965	9.925	22.328
..	...	...	...	...	...	...
..	...	...	...	...	...	...
100	1.290	1.660	1.984	2.364	2.626	3.174

90% Confidence Interval

$$\bar{x} \pm t^* \frac{s}{\sqrt{n}} = 13.5 \pm 1.660 * \frac{10.7}{\sqrt{145}} = 13.5 \pm 1.660 * 0.889 = 13.5 \pm 1.48$$

Adding and subtracting from 13.5 the margin of error of 1.48, we get a 90% confidence interval of (12.02 to 14.98). Table 6.6 offers annotated Minitab output of this interval.

Interpretation: We are 90% confident that the mean margin of victory for NFL home teams is from 12.02 to 14.98 points.

```
Variable      N    Mean   StDev  SE Mean        90% CI
Win Margin   145   13.50  10.70   0.889   (12.029,14.971)
              ↑      ↑      ↑       ↑               ↑
          Sample Sample Standard  Standard  Endpoints for 90%
           Size   Mean  Deviation  Error    Confidence Interval
```

TABLE 6.6: Annotated Minitab output for 95% confidence interval for Example 5—mean margin of victory for NFL home teams.

Example 6: How Far Do the Top College Recruits Travel in Selecting a College?

What is the mean driving distance between the ESPN Top 150 football recruits and their chosen college? Using the top 36 players listed as ESPN Top 150 recruits for 2012 and www.travelmath.com, the mean driving distance between the recruit's hometown and college was 439.4 miles, with a standard deviation of 542.4 miles. Calculate a 98% confidence interval for this mean.

As always, we check to verify that the t-methods can be applied in the calculation of the interval. With a sample size of 36, we can assume the sample mean will follow an approximately normal distribution based on the Central Limit Theorem. The t-methods would apply.

Next, we complete our confidence interval equation using 439.4 as the sample mean and 542.4 as the sample standard deviation. From the T-table, a portion of which is provided in Figure 6.5, the degrees of freedom (df) are 35. This gives us a t-multiplier 2.438 for a 98% confidence interval.

FIGURE 6.5: Portion of T-Table for 98% confidence interval in Example 6.

T-Table: t Distribution Confidence Interval and Critical Values

	Confidence Level					
	80%	90%	95%	98%	99%	99.8%
	Right Tail Probability					
df	$t_{0.10}$	$t_{0.05}$	$t_{0.025}$	$t_{0.01}$	$t_{0.005}$	$t_{0.001}$
1	3.078	6.314	12.706	31.821	63.657	318.289
2	1.886	2.920	4.303	6.965	9.925	22.328
..	...	...	...	...	...	...
..	...	...	...	...	...	...
35	1.306	1.690	2.030	2.438	2.724	3.340

98% Confidence Interval

$$\bar{x} \pm t^* \frac{s}{\sqrt{n}} = 439.4 \pm 2.438 * \frac{542.4}{\sqrt{36}} = 439.4 \pm 2.438 * 90.4 = 439.4 \pm 220.4$$

Adding and subtracting from 439.4 the margin of error of 220.4, we get a 98% confidence interval of (219.0 to 659.8). Table 6.7 gives annotate Minitab output for this calculation.

Interpretation: We are 98% confident that the mean driving distance between ESPN Top 150 football recruits and their chosen college is from 219 to 659.8 miles.

```
Variable        N   Mean   StDev   SE Mean       98% CI
Distance       36  439.4   542.4    90.4     (219.0, 659.8)
                ↑     ↑       ↑        ↑            ↑
           Sample Sample  Standard  Standard  Endpoints for 98%
             Size   Mean  Deviation  Error    Confidence Interval
```

TABLE 6.7: Annotated Minitab output for 98% confidence interval for Example 6—mean distance top college football recruits travel to selected college.

6.4 Finding Sample Size for Estimating a Population Mean

Calculating sample size for estimating a population mean is similar to that for estimating a population proportion: We solve for n in our margin for error. However, since the t-distribution is not as neat as the standard normal distribution, the process can be iterative. This means that we would solve, reset, solve, reset, etc., until we reached a conclusion. Yet, we can avoid this iterative process if we employ an approximate method based on the t-distribution, approaching the standard normal distribution as the sample size increases. This approximate method invokes the following formula:

$$n = \left(\frac{Z * \sigma}{M.E.}\right)^2$$

where σ is a population standard deviation possibly based on prior studies or knowledge, and Z comes from the z-multipliers in Table 6.1. Another possible way to estimate this standard deviation is to consider the Empirical Rule, where almost all observations fall within three standard deviations of the mean. Here, we could use an educated guess on the range of the data and divide by 4 to get a suitable value for σ.

Example: What sample size would be needed to estimate the mean recruiting distance in Example 3 if we wanted a margin of error of 300 miles with a 95% level of confidence? Before figuring this out, think of what sample size you would guess is required: maybe 20, or 50, possibly more? The longest driving distance in the continental United States is about 3500 miles from the southern tip of Florida to Washington State. Dividing 3500 by 4, we get a crude guess of 875 miles for σ. For a 95% level of confidence:

$$n = \left(\frac{1.96 * 875}{300}\right)^2 = 5.72^2 = 32.7 \text{ or } 33.$$

A surprisingly small sample! How was your guess?

Expressions and Formulas

1. Formula for a one-proportion confidence interval is:
$\hat{p} \pm Z * \sqrt{\frac{\hat{p}(1 - \hat{p})}{n}}$ where "Z" is a multiplier based on the level of confidence. Some common confidence levels and corresponding multipliers are:

Confidence	Z Multiplier
90%	1.65
95%	1.96
98%	2.33
99%	2.58

2. The margin of error for a one-proportion confidence interval is:

$$Z*\sqrt{\frac{\hat{p}(1-\hat{p})}{n}}$$

3. To find the sample size required for a specific margin of error and confidence level for a one-proportion confidence interval (use 0.5 as a conservative estimate of is one is not provided):

$$n=\frac{z^2\hat{p}(1-\hat{p})}{M.E.^2}$$

4. Formula for a one-mean confidence interval is:

$\bar{x}\pm t*\frac{s}{\sqrt{n}}$ where "t" is a t-multiplier from the T-table.

5. DF stands for "degrees of freedom" and is found by n - 1.

6. The margin of error for a one-mean confidence interval is:

$$t*\frac{s}{\sqrt{n}}$$

7. To find the sample size required for a specific margin of error and confidence level for a one-mean confidence interval:

$n=\left(\frac{Z*\sigma}{M.E.}\right)^2$ where "z" can come from the z-multipliers provided in No. 1 and σ is the population standard deviation which can be estimated by taking the range and dividing by 4.

Hypothesis Testing 7

In Chapter 6, we used confidence intervals to estimate some unknown population parameter. For example, we constructed one-proportion confidence intervals to estimate the proportion of NFL teams that win at home. However, there are occasions when our research leads us to testing some **hypothesis** about a population. This hypothesis is a statement about a population. The hypothesis is usually stated in terms of some parameter taking on a specific value. In trying to distinguish between a confidence interval and a hypothesis, consider the following two scenarios:

Scenario 1: You want to estimate the winning percentage of NFL teams playing at home.
Scenario 2: You want to research if there is a home field advantage in the NFL.

In Scenario 1, the goal is to find the range of winning percentages for home teams. The best statistical method to apply would be a confidence interval, where we could *estimate*—with some level of confidence—the winning percentage for NFL teams playing at home.

However, in Scenario 2, our interest has changed. Here, we are interested in *testing* if an advantage exists for NFL teams playing at home. The question is, what defines "an advantage?" Simply put, an advantage would occur if the home team won more games than they lost. In other words, a home field advantage would exist if the home team won more than 50% of the time. To answer this question, we apply **hypothesis testing** methods.

Hypothesis Testing—also called **significance testing**—is a statistical method where we summarize data to provide evidence regarding a hypothesis. Ultimately, we will take statistics from a sample (e.g., sample proportions and sample means) and use them to draw conclusions about unknown parameters of a population (e.g., population proportion and population mean). This process, using statistics to make judgments or decisions regarding population parameters, is another method of statistical inference called hypothesis testing.

Applying this definition to Scenario 2, we would take a random sample of NFL games, summarize the percentage of home teams that won, and compare this result to 50%. If our sample data provided evidence that the winning percentage for home teams exceeded 50%, we would have statistical support to conclude that there is an advantage to playing at home in the NFL.

As we did in Chapter 6 in terms of confidence intervals, we will begin by introducing hypothesis testing methods as they apply to one-proportion and one-mean situations. Similar to our discussion in that chapter, the key will be to recognize which parameter is being tested: that is, which hypothesis test to perform.

7.1 Steps to Performing a Hypothesis Test [1]

Step 1: State Hypotheses

In a significance test, we need to have two competing hypotheses: one called the **null hypothesis** and one called the **alternative** or **research hypothesis**. The key to remember is that our testing interests are about a population. This means that our notation must reflect that of the population. The **null hypothesis** will be a statement on what particular value the parameter equals, while the **alternative hypothesis** will be a statement on a particular range of values on which the parameter value can fall. Common notations used to denote these hypotheses are:

Ho for the **null hypothesis**

Ha for the **alternative hypothesis**

A general format for these hypotheses statements is provided below. Keep in mind that we compare only *two* hypotheses: One null hypothesis versus one alternative hypothesis. For simplicity, the null hypothesis will always include the "equal" sign, while the alternative will be chosen between "less than," "greater than," or "not equal," depending upon the research question of interest. The first two are referred to as *one-sided* tests, since our interest is in testing a specific direction against the null hypothesis, while the latter "not equal" is referred to as a *two-sided* test because the research interest is not directionally specific: We are just interested in showing the parameter is not some value. For instance, in Scenario 2 at the beginning of this chapter, the question was, "Does there exist a home field advantage in the NFL?" This implies a one-sided alternative hypothesis of "greater than 50%."

One Proportion (recall that "p" denotes population proportion)

$$Ho: p = p_0$$

$$Ha: p < p_0 \text{ or } Ha: p > p_0 \text{ or } Ha: p \neq p_0$$

Again, we pick only *one* of these alternative hypotheses, Ha, depending on the research question. When we conduct a hypothesis test, the "p_0" is replaced with a value such as 0.50 from Scenario 2. As an example of the proper notation for Scenario 2, the hypotheses statements would be written:

1 In discussing these steps, some books might use four steps, while others use five, as some steps might be combined. Regardless, the overall process remains the same.

$$Ho: p = 0.5$$

$$Ha: p > 0.5$$

One Mean (recall ← denotes population mean)

$$Ho: \mu = \mu_o$$

$$Ha: \mu < \mu_o \text{ or } Ha: \mu > \mu_o \text{ or } Ha: \mu \neq \mu_o$$

Again, we pick only *one* of these alternative hypothesis, Ha, depending on the research question. When we conduct a hypothesis test, the μ_o is replaced with a value. For example, NFL teams are considered to have an advantage of three points by playing at home (think of this as starting the game "up" three points). We want to research this claim to show that this is not true, although we are not sure if the advantage would be more or less than three points. The hypotheses statements would be set up as:

$$Ho: \mu = 3$$

$$Ha: \mu \neq 3$$

Step 2: Check Assumptions

First and foremost, hypothesis testing is based on the data representing a random sample taken from a particular population. Next, the data must follow certain distribution properties for that particular hypothesis test. With our introduction to hypothesis testing beginning with a discussion on tests of one proportion and one mean, we have the following assumptions:

One Proportion

$n*p_o$ and $n*(1-p_o)$ must be at least 15, where p_0 is the hypothesized proportion (e.g., 0.50 in the prior Scenario 2). If ***both*** conditions are satisfied, then we can employ Z-methods similar to what we used for one-proportion confidence intervals in Chapter 6.

One Mean

Either the population data is normal or approximately normal distributed, **or** the sample size is large enough: at least 30. This is the **Central Limit Theorem** introduced in Chapter 5.

Step 3: Set a Level of Significance (Denoted by the Greek Symbol alpha, α)

The level of significance refers to a "baseline level" of probability. After gathering our data, we will calculate the probability we would get such data as it relates to our null hypothesis; that is, we will calculate the probability our sample data produced a particular result under the assumption that the null hypothesis was true. We will compare this sample data probability to the level of significance to make a statistical decision regarding the null hypothesis. The most common level of significance[2] is 0.05, or 5%. This is the significance level we will use for the remainder of this text whenever we conduct a significance test.

Step 4: Calculate Test Statistic

When we gather our data and calculate a statistic, a natural inclination would be to compare that statistic to the parameter. As we learned in Chapter 6, we expect some error in our sampling; we would not expect our statistic to be *exactly* equal to the parameter, even if the parameter were true. However, we would expect the statistic to be close to the parameter *if* the parameter (think null hypothesis!) were true. For instance, if NFL home teams won on average by three points, we would not expect our sample data to have a difference of exactly three points. Yet if this three point difference was true, we would expect the mean of our sample to be close to three points. A test statistic is a comparison of the sample statistic to the parameter while allowing for some measure of error, under the assumption that the null hypothesis is true. If the assumptions (see Step 1) are satisfied, we calculate the test statistics as shown below. The use of the subscript "stat" is to clarify that the number calculated refers to a test statistic. As a helpful hint, think of these test statistics as puzzles, where you need to insert the numbers into their respective place in the puzzle. A common error for students is to want to work with larger numbers first. This leads them to always use in the numerator the larger value minus the smaller value. Avoid that practice! If you compare both test statistics, you will see the same general format in the numerator: *sample statistic minus parameter*.

One Proportion

$$Z_{stat} = \frac{\hat{p} - p_o}{\sqrt{\frac{p_o(1 - p_o)}{n}}}$$

One Mean

$$t_{stat} = \frac{\bar{x} - \mu_o}{\frac{s}{\sqrt{n}}}$$

2 Some other common levels are 0.1 (10%) and 0.01 (1%).

Step 5: Calculate p-value[3]

The term "p-value" is short for "probability value," and therefore is a probability and must follow the basic properties of probability, such as the p-value must fall somewhere from zero to one, inclusive. This p-value is defined as,

The probability our sample data produced this result given that the null hypothesis is true.

In thinking of the p-value, keep in mind that we begin by assuming the null hypothesis is correct: NFL home teams only win half the time or the home team wins on average by three points. As stated earlier in Step 3, when we gather the sample data, we expect the sample statistic to differ from the parameter. The question becomes how unlikely was it for us to get this difference. If the statistic and parameter are "close," this would not be surprising—remember, we are assuming the null hypothesis (parameter) is true—and therefore have a high probability (p-value) of happening. However, if the statistic and parameter are far apart, this would be surprising and lead us to question the null hypothesis. Again, since we assume the null hypothesis to be true, the chances would be unlikely (small p-value) that we would get sample data that produces a statistic much different from the assumed parameter value.

Consider the NFL home field advantage scenario: We begin by assuming no home field advantage, then we would expect our sample data to produce a sample proportion near 0.5 or 50%. Let's say four independent students are researching this hypothesis by going out and taking a random sample of 100 NFL games. These four studies produce the following sample proportion of wins for the home team:

Study 1: $\hat{p}$= 0.51

Study 2: $\hat{p}$= 0.53

Study 3: $\hat{p}$= 0.61

Study 4: $\hat{p}$= 0.65

Which of these studies do you believe would provide the highest probability of happening *if* 0.5 were the true proportion? Which of these studies do you believe would provide the smallest probability of happening *if* 0.5 were the true proportion? Which result would be more convincing in going against the null hypothesis?

Study 1 with the sample proportion closest to 0.5 would be the most likely (highest p-value) and Study 4 with sample proportion furthest from 0.5 would be the least likely (smallest p-value), assuming 0.5 was the true winning proportion for NFL teams playing at home.

As a sample statistic moves further away from the null value, we would begin to question this null hypothesis because getting such a difference would be unlikely if the null hypothesis were true. Therefore, the smaller the p-value (the more unlikely the sample result), the stronger the evidence we have against the null hypothesis.

3 Some introductory statistics courses and classes may also use a rejection region approach to hypothesis testing. We explain only p-value approach, as this is most commonly used in practice (e.g., research journals).

In typical applications, statistical software is used to perform hypothesis tests and calculate the p-value. However, for educational purposes, students might be asked to perform these tests by hand, including the calculation of the p-value. We will introduce a method for finding the p-value for a hypothesis test now, prior to moving on to Step 6.

Calculating p-value by Hand Using Z- and T-Tables

One important aspect of satisfying the assumptions for a significance test, Step 1, is that by doing so one can use the Z- and T-Tables to find p-values after calculating the test statistic. Calculation of the p-value will be based on the choice of alternative hypothesis.

One Proportion (Z-Table)

Ha is less than: If the alternative hypothesis is the one-sided "less than," the p-value is the probability from the Z-Table for getting less than the Z_{stat}. That is, we go to the Z-Table, look up the Z_{stat}, and the cumulative probability for this Z_{stat} is the p-value.

Ha is greater than: If the alternative hypothesis is the one-sided "greater than," the p-value is the probability from the Z-Table for getting more than the Z_{stat}. That is, we go to the Z-Table, look up the Z_{stat}, take the cumulative probability for this Z_{stat} and **subtract this cumulative probability from one** to get the p-value.

Ha is not equal: If the alternative hypothesis is the two-sided "not equal," the p-value is twice the probability from the Z-Table for getting more than the absolute value of the Z_{stat}. That is, we take the absolute value of the Z_{stat}, take the cumulative probability for this Z_{stat} from the Z-Table, subtract this cumulative probability from one. We then double this to get the p-value.

One Mean (T-Table)

With the t-distribution being symmetrical and including the degrees of freedom, this information is used to create a T-Table that provides the right-tail probabilities for combinations of degrees of freedom and T values. The right-tail probability is the probability, or area, to the right of the absolute value of a T value. This probability would be similar to taking one minus the cumulative probability for a T value. Due to this, the T-Table will be used to get a range for the p-value instead of one specific p-value.

Ha is less than or greater: If the alternative hypothesis is either the one-sided "less than" or the one-sided "greater than," the p-value is found by:

- Getting the degrees for freedom by n - 1, where "n" is the sample size.
- Going across that DF row in the T-Table locating where the t_{stat} would fall (getting a t_{stat} that *exactly* matches one in the table is extremely unlikely!).

- After locating next to what T-values the t_{stat} would fall, find the right-tail probabilities associated with those T-values.
- The p-value will fall between these right-tail probabilities. That is, the T-Table provides a *range* for possible p-values.

Ha is not equal: If the alternative hypothesis is the two-sided "not equal," the p-value is found by following the previous steps and then doubling the endpoints found in Step 4.

Here are some examples for reading the T-Table to find p-values - Figure 7.1 provides a portion of the T-table:

- You have a t_{stat} = 2.11 and a sample of size 15 resulting in DF = 14: Go to the T-Table and the row for df of 14. Start going across that row to find where the t_{stat} of 2.11 would fall. From the t-table, this t_{stat} falls between 1.761 and 2.145, which have right-tail probabilities of 0.05 and 0.025, respectively (see the **red** numbers in Figure 7.1). Since the t_{stat} of 2.11 falls between these T-values, the right-tail probability must fall between 0.025 and 0.05 as well. Therefore, we would say the p-value falls between 0.025 and 0.05 or 0.025 < *p-value* < 0.05. If the alternative hypothesis is "not equal," we would simply double this range to say the p-value is between 0.05 and 0.10.
- You have a t_{stat} = -2.11 and a sample of size 15 resulting in DF = 14; you would take the absolute value (i.e., t_{stat} = 2.11) and follow the same steps outlined above.
- You have a t_{stat} = 4.35 and a sample of size 10 resulting in DF = 9: This t_{stat} is beyond the last T-value listed on the table for df of 9, the T-value of 4.297 (see the **blue** number in Figure 7.1). Since the right-tail probabilities *decrease* as you go left to right, this would imply that the right-tail probability for a t_{stat} of 4.35 is less than the right-tail probability for 4.297 T-value. Thus, the p-value would be less than 0.001. If the alternative hypothesis is not equal, we would double this value to say the p-value is less than 0.002.
- You have a t_{stat} = 1.20 and a sample size of 20 resulting in DF = 19: Go to the T-Table and find the row for df of 19. The first T-value we find is 1.328 (see the **black** number in Figure 7.1), which is greater than our t_{stat} of 1.20. With the right-tail probability of 0.100 for T-value of 1.328 and our t_{stat} being smaller than this T-value, this would mean that the right-tail probability for our t_{stat} would be greater than 0.100, or the p-value would be more than 0.100. Again, if the alternative hypothesis is "not equal, we would double this value to conclude that the p-value is greater than 0.200.

FIGURE 7.1: Portion of T-Table for finding p-values.

T-Table: t Distribution Confidence Interval and Critical Values

	Confidence Level					
	80%	90%	95%	98%	99%	99.8%
	Right Tail Probability					
df	$t_{0.10}$	$t_{0.05}$	$t_{0.025}$	$t_{0.01}$	$t_{0.005}$	$t_{0.001}$
1	3.078	6.314	12.706	31.821	63.657	318.289
2	1.886	2.920	4.303	6.965	9.925	22.328
..	...	...	...	...	...	...
9	1.383	1.833	2.262	2.821	3.250	4.297
..	...	...	...	...	...	...
14	1.345	1.761	2.145	2.624	2.977	3.787
..	...	...	...	...	...	...
19	**1.328**	1.729	2.093	2.539	2.861	3.579

Step 6: Making a Decision and Drawing a Conclusion

The conclusion for a hypothesis test involves making a decision regarding the null hypothesis, then summarizing the study results. The decision is straightforward:

If the p-value is less than the level of significance (0.05 for our purposes), we reject the null hypothesis. However, if the p-value is greater than 0.05, we fail to reject the null hypothesis.[4]

The conclusion would be a recap of the test, including the p-value, decision, and what this decision means in terms of the null hypothesis. Notice that the decision is in terms of rejecting or failing to reject the *null* hypothesis. Recall we started by assuming this null hypothesis was correct until we have statistical evidence to say otherwise. A **statistically significant result** is one where the null hypothesis is rejected.

7.2 Hypothesis Testing: One Proportion

Example 1: NFL Home Field Advantage

Is there a home field advantage playing in the NFL? Using the results of the 2011 NFL season as a sample of all NFL games, we find that the home team won 145 of 256 games played.

4 We avoid the conflict of "equal to," since the chances of this happening are extremely unlikely and would most likely occur due to rounding error of our calculations.

Step 1: State Hypotheses

Since the question regards a test of proportion, we must use parameter notation in terms of proportions. With the research question being "Is there an advantage to playing at home in the NFL?" the implication is that we want to show that significantly more than a majority of the time the home team wins. This sets the hypotheses as:

$$Ho: p = 0.5 \text{ versus } Ha: p > 0.5$$

Step 2: Check Assumptions

$n*p_0$ and $n*(1-p_0)$ must be at least 15, where here p_0 is 0.5 and n is 256. The calculations, 256*0.5 = 128 and 256*(1 – 0.5) = 128, both exceeding 15 and satisfying the assumption.

Step 3: Set Level of Significance

As previously stated, we will use 0.05 for all hypothesis tests in this book.

Step 4: Calculate Test Statistic

The sample proportion is 145/256 = 0.57 and substituting appropriately into:

$$\backslash Z_{stat} = \frac{\hat{p} - p_o}{\sqrt{\frac{p_o(1-p_o)}{n}}} = Z_{stat} = \frac{0.566 - 0.5}{\sqrt{\frac{0.5(1-0.5)}{256}}} = \frac{0.066}{0.03125} = 2.11$$

Step 5: Calculate p-value

With the alternative hypothesis being "greater than," we get the p-value by finding the probability of getting greater than the Z_{stat} of 2.11 in the Z-Table, then subtracting this cumulative probability from one. From the Z-Table, a portion of which is provided in Figure 7.2, we find the cumulative probability for 2.11 is 0.9826, leading to a p-value of 0.0174.

FIGURE 7.2: Portion of Z-Table for Example 1.

Z-Table (continued): Standard Normal Cumulative Probabilities

Cumulative probability (area to LEFT) of Positive Z-values

Z	0.00	0.01	0.02	0.03	0.04	0.05	0.06	0.07	0.08	0.09
0.0	0.5000	0.5040	0.5080	0.5120	0.5160	0.5199	0.5239	0.5279	0.5319	0.5359
0.1	0.5398	0.5438	0.5478	0.5517	0.5557	0.5596	0.5636	0.5675	0.5714	0.5753
0.2	0.5793	0.5832	0.5871	0.5910	0.5948	0.5987	0.6026	0.6064	0.6103	0.6141
...	...	...	...	...	...	...	...	...	...	...
...	...	...	...	...	...	...	...	...	...	...
2.1	0.9821	0.9826	0.9830	0.9834	0.9838	0.9842	0.9846	0.9850	0.9854	0.9857

Step 6: Conclusion

Comparing the p-value of 0.0174 to 0.05, we find the p-value to be smaller. Therefore, we reject the null hypothesis in favor of the alternative. We would conclude, based on our sample data of NFL games played in 2011, that there is a statistically significant advantage to playing at home in the NFL. Annotated Minitab output is provided in Table 7.1.

In effect, we are saying that the sample proportion of 56.6% was too different from the hypothesized proportion of 50% to be due to random chance alone. The probability that our sample data produced a sample proportion this far, or further, from the hypothesized value of 50% was extremely unlikely. This leads one to conclude that the null hypothesis is incorrect and that the true population proportion for home team winning in the NFL is greater than 50%. A hypothesis test does NOT tell us what the population proportion is—that is, we cannot say the actual proportion is 56.6%. Just that we have statistical evidence to state that the population proportion is greater than 50%. A confidence interval would help us estimate what range we believe the true population proportion falls. This general reasoning will be true for all hypothesis tests.

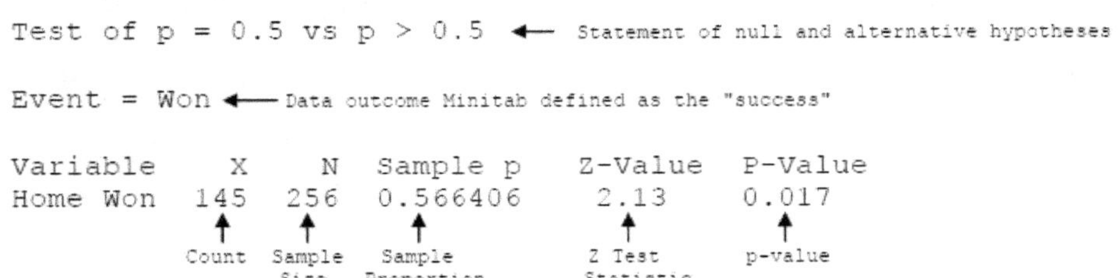

```
Test of p = 0.5 vs p > 0.5  ←— Statement of null and alternative hypotheses

Event = Won ←— Data outcome Minitab defined as the "success"

Variable    X    N   Sample p   Z-Value   P-Value
Home Won  145  256   0.566406     2.13     0.017
           ↑    ↑       ↑           ↑         ↑
        Count Sample  Sample     Z Test    p-value
               Size  Proportion Statistic
```

TABLE 7.1: Annotated Minitab output for Example 1—hypothesis test of one proportion.

Example 2: College Football Home Field Advantage

Is there a home field advantage in college football?[5] Using the results of the 2011 college season (Division 1 or FBS) as a sample of all Division 1 college games, we find that the home team won 483 of 748 games played.

Step 1: State Hypotheses

Since the question regards a test of proportion, we need to use parameter notation in terms of proportions. With the research question, "Is there an advantage to playing at home in college football?" the implication is that we want to show that significantly more than a majority of the time the home team wins. This sets the hypotheses as:

$$Ho: p = 0.5 \text{ versus } Ha: p > 0.5$$

Step 2: Check Assumptions

$n*p_0$ and $n*(1-p_0)$ must be at least 15, where here p_0 is 0.5 and n is 748. The calculations, 748*0.5 = 374 and 748*(1-0.5) = 374, both exceeding 15 and satisfying the assumption.

Step 3: Set Level of Significance

As previously stated, we will use 0.05 for all hypothesis tests in this book.

Step 4: Calculate Test Statistic

The sample proportion is 483/748 = 0.646 and substituting appropriately into:

$$Z_{stat} = \frac{\hat{p} - p_o}{\sqrt{\frac{p_o(1 - p_o)}{n}}} = Z_{stat} = \frac{0.646 - 0.5}{\sqrt{\frac{0.5(1 - 0.5)}{748}}} = \frac{0.15}{0.0183} = 7.98$$

Step 5: Calculate p-value

With the alternative hypothesis being "greater than," we get the p-value by finding the probability of getting greater than the Z_{stat} of 7.98 in the Z-Table, then subtracting this cumulative probability from

5 We've kept with these two examples, as they present excellent segues into a topic covered in Chapter 8.

one. From the Z-Table, a portion of which is provided in Figure 7.3, the largest z-score is 3.09 with a cumulative probability of 0.9990, or p-value of 0.001. Since our Z_{stat} is beyond 3.09, it stands to reason that the cumulative probability for 7.98 would be more than 0.9990, meaning our p-value for this test would be less than 0.001.

FIGURE 7.3: Portion of Z-Table for Example 2.

Z-Table (continued): Standard Normal Cumulative Probabilities

Cumulative probability (area to LEFT) of Positive Z-values

Z	0.00	0.01	0.02	0.03	0.04	0.05	0.06	0.07	0.08	0.09
0.0	0.5000	0.5040	0.5080	0.5120	0.5160	0.5199	0.5239	0.5279	0.5319	0.5359
0.1	0.5398	0.5438	0.5478	0.5517	0.5557	0.5596	0.5636	0.5675	0.5714	0.5753
0.2	0.5793	0.5832	0.5871	0.5910	0.5948	0.5987	0.6026	0.6064	0.6103	0.6141
...	...	...	...	...	...	...	...	...	...	...
...	...	...	...	...	...	...	...	...	...	...
3.0	0.9987	0.9987	0.9987	0.9988	0.9988	0.9989	0.9989	0.9989	0.9990	0.9990

Step 6: Conclusion

With the p-value for our test statistic being less than 0.001, this would also be less than 0.05: Therefore, we reject the null hypothesis in favor of the alternative. We would conclude, based on our sample data of college games played in 2011, that there is a statistically significant advantage to playing at home. Annotated Minitab output is provided in Table 7.2.

Similar to Example 1, we are saying that the sample proportion of 64.6% was too different from the hypothesized proportion of 50% to be due to random chance alone. The probability that our sample data produced a sample proportion this far from the hypothesized value of 50% was extremely unlikely. This leads one to conclude that the null hypothesis is incorrect and that the true population proportion for home team winning in major college football is greater than 50%. A hypothesis test does NOT tell us what the population proportion is—that is, we cannot say the actual proportion is 64.6%. Just that we have statistical evidence to state that the population proportion is higher than 50%. We would need a confidence interval to estimate what range we believe the true population proportion falls.

```
Test of p = 0.5 vs p > 0.5  ←  Statement of null and alternative hypotheses

Event = Won ← Data outcome Minitab defined as the "success"

Variable    X    N    Sample p    Z-Value   P-Value
Home Won   483  748   0.645722     7.97      0.000
            ↑    ↑       ↑           ↑         ↑
          Count Sample  Sample     Z Test    p-value
                Size    Proportion Statistic
```

TABLE 7.2: Annotated SPSS output for Example 2—hypothesis test of one proportion.

Example 3: Baseball Average in Major League Baseball

Think of upcoming playoffs for any sport and conversations you may have had regarding them. Maybe you or a friend has said something along the lines of, "You can throw out regular season stats; they don't mean a thing now." Or possibly the phrase, "The postseason is a whole new ballgame." These remarks imply that what happens in the postseason differs from that of the regular season. How would this comment fit the St. Louis Cardinals team batting average during their championship run in 2011? We could answer by testing if their team batting average in the postseason differed significantly from their team batting average for the regular season. In 2011, the Cardinals as a team batted 0.273 in the regular season and 0.271 during the postseason by getting 167 hits in 616 at-bats (see www.mlb.com).

Step 1: State Hypotheses

Since the question regards a test of proportion, we need to use parameter notation in terms of proportions. With the research question, "Is there evidence that batting average in the postseason differs significantly from the regular season in MLB?" the implication is that we want to show that the Cardinals had a postseason batting average *different* from 0.273. This sets the hypotheses as:

$$\text{Ho: } p = 0.273 \text{ versus Ha: } p \neq 0.273$$

Step 2: Check Assumptions

$n*p_0$ and $n*(1-p_0)$ must be at least 15, where here p_0 is 0.273 and n is 616. The calculations, 616*0.273 = 168 and 616*(1-0.273) = 448, both exceeding 15 and satisfying the assumption.

Step 3: Set Level of Significance

As previously stated, we will use 0.05 for all hypothesis tests in this book.

Step 4: Calculate Test Statistic

The sample proportion is 167/616 = 0.271 and substituting appropriately into:

$$Z_{stat} = \frac{\hat{p} - p_o}{\sqrt{\frac{p_o(1-p_o)}{n}}} = Z_{stat} = \frac{0.271 - 0.273}{\sqrt{\frac{0.273(1-0.273)}{616}}} = \frac{-0.002}{0.018} = -0.11$$

Step 5: Calculate p-value

With the alternative hypothesis being "not equal," we get the p-value by doubling the probability of getting greater than the $|Z_{stat}|$ or 0.11 in the Z-Table. From the Z-Table, a portion of which is provided in Figure 7.4, the cumulative probability for 0.11 is 0.5438 making the probability 0.4562 of getting greater than 0.11. Finally, since the alternative is "not equal," we double this probability to arrive at the p-value of 0.9124 for this test.

FIGURE 7.4: Portion of Z-Table for Example 3.

Z-Table (continued): Standard Normal Cumulative Probabilities

Cumulative probability (area to LEFT) of Positive Z-values

z	0.00	0.01	0.02	0.03	0.04	0.05	0.06	0.07	0.08	0.09
0.0	0.5000	0.5040	0.5080	0.5120	0.5160	0.5199	0.5239	0.5279	0.5319	0.5359
0.1	0.5398	0.5438	0.5478	0.5517	0.5557	0.5596	0.5636	0.5675	0.5714	0.5753
0.2	0.5793	0.5832	0.5871	0.5910	0.5948	0.5987	0.6026	0.6064	0.6103	0.6141

Step 6: Conclusion

With the p-value for our test statistic being greater than 0.05, we fail to reject the null. We would conclude, based on the Cardinals' 2011 postseason, that there is not enough statistical evidence to support stating that what happens in the postseason differs from the regular season—at least in terms of team batting average in Major League Baseball. Annotated Minitab output is provided in Table 7.3. Additionally, this output also includes a 95% confidence interval to demonstrate how the concept of confidence intervals from Chapter 6 can be employed to support significance test conclusions. From this interval, we conclude that we are 95% confident that the Cardinals' play-off batting average is from 0.236 to 0.306. How does this interval support our decision to not reject the null hypothesis? Since our hypothesized value of 0.273 is contained within the confidence interval, it is plausible that the play-off batting average is 0.273, supporting our decision to not reject Ho.

```
Test of p = 0.273 vs p not = 0.273 ←Statement of hypotheses

Event = Hits ← Data outcome Minitab defined as the "success"

Variable       X    N   Sample p        95% CI        Z-Value  P-Value
Batting Avg   167  616  0.271104   (0.236,0.306)     -0.11     0.916
               ↑    ↑      ↑              ↑              ↑         ↑
            Count Sample  Sample    95% Confidence   Z Test    p-value
                   Size  Proportion    Interval     Statistic
```

TABLE 7.3: Annotated Minitab output, including a 95% confidence interval for Example 1—hypothesis test of one proportion.

7.3 Hypothesis Testing: One Mean

Example 4: Putting on PGA Tour

On the PGA tour, the difference between losing and winning often comes down to putting; the better one putts the better one's chances of winning. If each player reached the green in regulation (e.g. on a par 4 they reached the green in 2 shots) then to break par, a player would have to have fewer than 36 putts - based on 2 putts per 18 holes. However, since even the best golfers in the world are not perfect - they do not always reach the green in regulation - they are often forced to make only one putt on a green in order to keep pace. Based on data from 2003 through 2001 found at www.espn.go.com/golf, the average number of putts per round for a professional male golfer is roughly 32. As a research question, we are interested to determine if the top 40 putters on the men's PGA tour average less than 32 putts per round. Using statistics on the top 40 putters in 2011 as a random sample, the average number of putts per round was 31.4 with a 0.26 standard deviation.

Step 1: State Hypotheses

Since the question regards a test of mean—think of the data "putts" as quantitative, and therefore our interest is a mean—we need to use parameter notation in terms of means. With the research question, "Do the top 40 PGA tour putters average fewer than 32 putts per round?" the implication is that we want to show that the average number of putts per round for the Top 40 putters is less than 32. We set the hypotheses as:

$$Ho: \mu = 32 \text{ versus } Ha: \mu < 32$$

Step 2: Check Assumptions

The assumption that the sample mean is normal can be satisfied in one of two ways: The population data is approximately normal or the sample size is large enough—at least 30 (recall the Central Limit Theorem). With the sample size of 40, we could invoke the Central Limit Theorem.

Step 3: Set Level of Significance

As previously stated, we will use 0.05 for all hypothesis tests in this book.

Step 4: Calculate Test Statistic

The sample mean $\bar{X}$, is 31.4 and standard deviation, S, is 0.26. Substituting appropriately into the test statistic equation:

$$t_{stat} = \frac{\bar{X} - \mu_o}{\frac{S}{\sqrt{n}}} = \frac{31.4 - 32}{\frac{0.26}{\sqrt{40}}} = \frac{-0.6}{0.041} = -14.63$$

Step 5: Calculate p-value

We begin with the degrees of freedom, DF, equal to n – 1, which for this example comes to 39. Using the absolute value of our t_{stat} or 14.63, we search the T-table, finding where 14.63 falls within the DF row of 39, or the closest DF to 39 without exceeding. From the T-table, a portion of which is provided in Figure 7.5, the closest DF to 39 without exceeding is 35. Going across this row, we see that the values increase ending at 3.340. This 3.340 corresponds to a right-tail probability of 0.001. That is, with a DF of 35, there is a 0.001 probability of getting a t-statistic of 3.340 or greater. With our t_{stat} of 14.63 exceeding 3.340, there is less than a 0.001 probability of getting a t-statistic of 14.63 or greater. With the alternative hypothesis being "less than," this gives us our p-value for the test. The p-value would be less than 0.001.

FIGURE 7.5: Portion of T-Table for Example 4.

T-Table: t Distribution Confidence Interval and Critical Values

	Confidence Level					
	80%	90%	95%	98%	99%	99.8%
	Right Tail Probability					
df	$t_{0.10}$	$t_{0.05}$	$t_{0.025}$	$t_{0.01}$	$t_{0.005}$	$t_{0.001}$
1	3.078	6.314	12.706	31.821	63.657	318.289
2	1.886	2.920	4.303	6.965	9.925	22.328
..	...	...	...	...	...	...
..	...	...	...	...	...	...
35	1.306	1.690	2.030	2.438	2.724	3.340

Step 6: Conclusion

With the p-value being less than 0.001, we know that the p-value is also less than the 0.05 level of significance. Therefore, we reject the null hypothesis in favor of the alternative. We would conclude, based on our sample data of the 2011 PGA tour, that there is statistically significant evidence to conclude that the top 40 putters on the men's PGA tour average fewer than 32 putts per round. Annotated Minitab output is provided in Table 7.4. As we see in this output the p-value of 0.000 is in line we our calculations of the p-value being less than 0.001 from the T-Table.

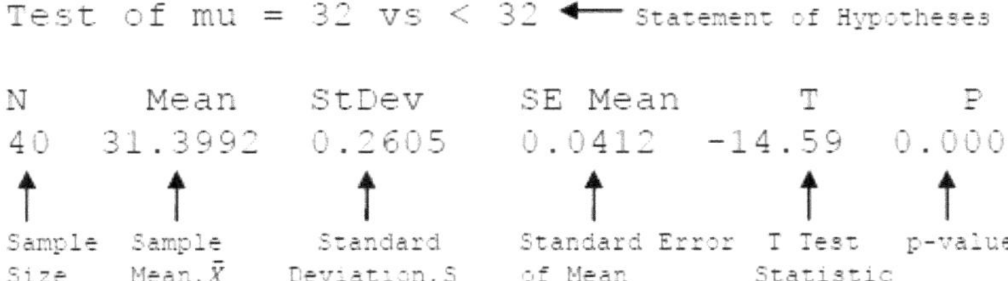

```
Test of mu = 32 vs < 32  <-- Statement of Hypotheses

N      Mean     StDev    SE Mean      T        P
40   31.3992   0.2605    0.0412    -14.59    0.000
```

Sample Size	Sample Mean, $\bar{X}$	Standard Deviation, S	Standard Error of Mean	T Test Statistic	p-value
40	31.3992	0.2605	0.0412	-14.59	0.000

TABLE 7.4: Annotated Minitab output for Example 4—hypothesis test of one mean using 2011 PGA tour data.

In effect, we are saying that the sample mean of 31.4 was too different from the hypothesized mean of 32 to be due to random chance alone. The probability that our sample data produced a sample mean this far, or further, from the hypothesized value of 32 was extremely unlikely. Thus, we conclude that the null hypothesis is incorrect and that the mean number of putts per round for the top 40 putters on the men's PGA tour is less than 32. The hypothesis test does NOT tell us what the true mean is—that is we cannot say the true population mean is 31.4, just that we have statistical evidence to state that the population mean number of putts per round for the top 40 putters is less than 32. A confidence interval would help us estimate in what range we believe the parameter falls. This general reasoning will be true for all hypothesis tests.

Example 5: Target Bench Press Numbers at NFL Combine

According to NFL.com, the target bench press number for interior linemen—offensive guards and centers—is 26 reps of 225 pounds. Using the 2012 combine results as a sample of all prospective NFL interior linemen, is there evidence to suggest that 26 reps is not reasonable? In other words, is the average number of reps for interior linemen something other than 26? Using data located at www.footballsfuture.com, the mean reps for 17 interior linemen who participated in the 2012 NFL combine was 26.71, with a standard deviation of 6.59.

Step 1: State Hypotheses

Since the question regards a test of mean, we need to use parameter notation in terms of means. With the research question, "Is there evidence to suggest that the average bench press reps for interior linemen is not 26?" we set the hypotheses as:

$$\text{Ho: } \mu = 26 \text{ versus Ha: } \mu \neq 26$$

Step 2: Check Assumptions

The assumption that the sample mean is normal can be satisfied in one of two ways: The population data is approximately normal or the sample size is large enough—at least 30 (recall the Central Limit Theorem). Since our data size is small—less than 30—we cannot apply the Central Limit Theorem. We also do not know the population distribution of bench press data, so we are left to examine the shape of the sample data. A common graphing option is a **probability plot**, which plots the data against a specified distribution, e.g., a normal distribution. The goal is for the probability plot to provide evidence in support of a normal distribution. This decision is demonstrated with the plotted values falling close to a linear relationship or if the p-value for a test of normality is greater than 0.05. From the probability plot shown in Figure 7.6, we can see that normality can be assumed. The plotted points fall within the 95% confidence bands of a straight line and the p-value, 0.766, exceeds 0.05.

FIGURE 7.6: Probability plot of interior linemen reps of 225 pounds bench press at 2012 NFL combine.

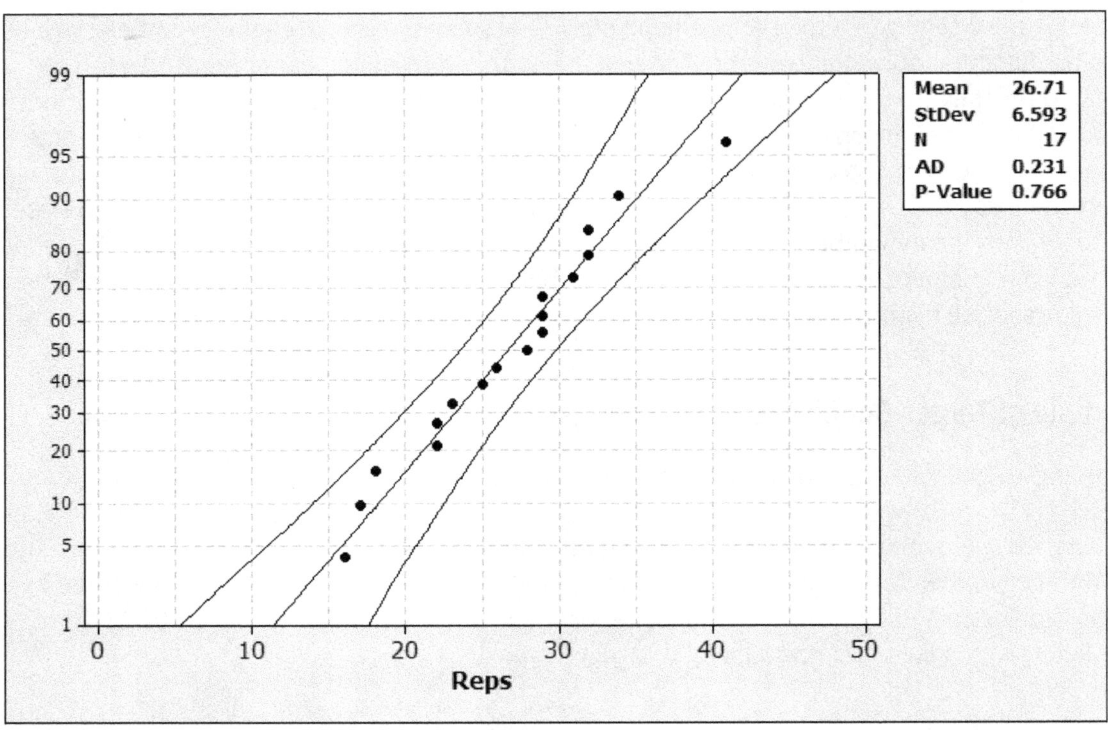

Step 3: Set Level of Significance

As previously stated, we will use 0.05 for all hypothesis tests in this book.

Step 4: Calculate Test Statistic

The sample mean, $\bar{X}$, is 26.71 and standard deviation, S, is 6.59. Substituting appropriately into the test statistic equation: $t_{stat} = \frac{\bar{X} - \mu_o}{\frac{S}{\sqrt{n}}} = \frac{26.71 - 26}{\frac{6.59}{\sqrt{17}}} = \frac{0.71}{1.60} = 0.44$

Step 5: Calculate p-value

We begin with the degrees of freedom, DF, equal to n − 1, which for this example comes to 16. Using the absolute value of our t_{stat} or 0.44, we search the T-Table, a portion of the T-Table is provided in Figure 7.7, finding where 0.44 falls within the DF row of 16. Starting in that row, the first T-statistic we come to is 1.337, which corresponds to a right-tail probability of 0.10. That is, with a DF of 16, there is a 0.10 probability of getting a t-statistic of 1.337 or greater. With our t_{stat} of 0.44 preceding 1.337, there is more than a 0.10 probability of getting a t-statistic of 0.44 or greater. With the alternative hypothesis being "not equal," we must double this right-tail probability. This results in the p-value for our test being greater than 0.20.

FIGURE 7.7: Portion of T-Table for Example 5.

T-Table: t Distribution Confidence Interval and Critical Values

	Confidence Level					
	80%	90%	95%	98%	99%	99.8%
	Right Tail Probability					
df	$t_{0.10}$	$t_{0.05}$	$t_{0.025}$	$t_{0.01}$	$t_{0.005}$	$t_{0.001}$
1	3.078	6.314	12.706	31.821	63.657	318.289
2	1.886	2.920	4.303	6.965	9.925	22.328
..	...	...	...	...	...	...
..	...	...	...	...	...	...
16	1.337	1.746	2.120	2.583	2.921	3.686

Step 6: Conclusion

With the p-value being greater than 0.20, we know that the p-value is also more than the 0.05 level of significance. Therefore, we fail to reject the null hypothesis. We would conclude, based on our sample data of the 2012 NFL combine, that there is not enough statistical evidence to conclude that the average number of bench press reps for interior linemen differs from 26. This would suggest that 26 reps is a reasonable target number for interior linemen. However, this does not *confirm* the null hypothesis, nor should one interpret this to mean the null hypothesis is true. By failing to reject the null hypothesis, we are saying there was not enough evidence to reject the null hypothesis. Annotated Minitab output is provided in Table 7.5. Notice that Minitab has provided a more exact p-value of 0.665 for this test. This value agrees with our p-value being greater than 0.20 when using the T-Table. The output also includes a 95% confidence interval for estimating the mean number of bench press reps. This interval can be interpreted as "We are 95% confident that the mean number of bench press reps completed by interior linemen at the NFL combine is from 23.32 to 30.10." The interval supports our decision to not reject the null hypothesis, since the hypothesized value of 26 is captured by this interval.

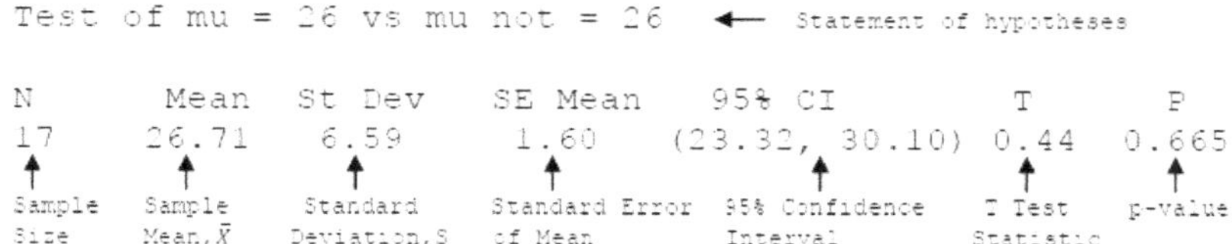

```
Test of mu = 26 vs mu not = 26   ← Statement of hypotheses

N      Mean   St Dev   SE Mean       95% CI          T       P
17    26.71    6.59     1.60    (23.32, 30.10)    0.44   0.665
```

Sample Size	Sample Mean, $\bar{X}$	Standard Deviation, S	Standard Error of Mean	95% Confidence Interval	T Test Statistic	p-value

TABLE 7.5: Annotated Minitab output for Example 5—hypothesis test of one mean using 2012 NFL combine data.

Example 6: Is Colorado's Coors Field a Home Run Park?

An oft-cited comment regarding Coors Field, home of the Colorado Rockies, is that it is a hitter's ballpark, frequently resulting in more home runs than the average ballpark. Using data from 2000 through 2011 to represent sample data (see www.espn.go.com/mlb), is there evidence to support that Coors Field surrenders more home runs per game than the average ballpark? During this time frame, the average home runs were 1.55 per stadium, not including Coors Field. Over these 12 years, Coors Field surrendered an average of 1.77 home runs, with a 0.49 standard deviation.

Step 1: State Hypotheses

Since the question regards a test of mean, we need to use parameter notation in terms of means. With the research question, "Does Coors Field surrender, on average, more home runs than the other major league ballparks?" we set the hypotheses as:

$$Ho: \mu = 1.55 \text{ versus } Ha: \mu > 1.55$$

Step 2: Check Assumptions

The assumption that the sample mean is normal can be satisfied in one of two ways: The population data is approximately normal or the sample size is large enough—at least 30 (recall the Central Limit Theorem). Since our data size is small—less than 30—we cannot apply the Central Limit Theorem. We also do not know the population distribution of bench press data, so we are left to examine the shape of the sample data. A common graphing option is a **probability plot**, which plots the data against a specified distribution, e.g., a normal distribution. The goal is for the probability plot to provide evidence in support of a normal distribution. This decision is demonstrated with the plotted values falling close to a linear relationship or if the p-value for a test of normality is greater than 0.05. From the probability plot shown in Figure 7.8, we can see that normality can be assumed—barely! The plotted points fall within the 95% confidence bands of a straight line and the p-value, 0.059, exceeds 0.05.

FIGURE 7.8: Probability plot of home runs per game hit at Coors Field from 2000 to 2011.

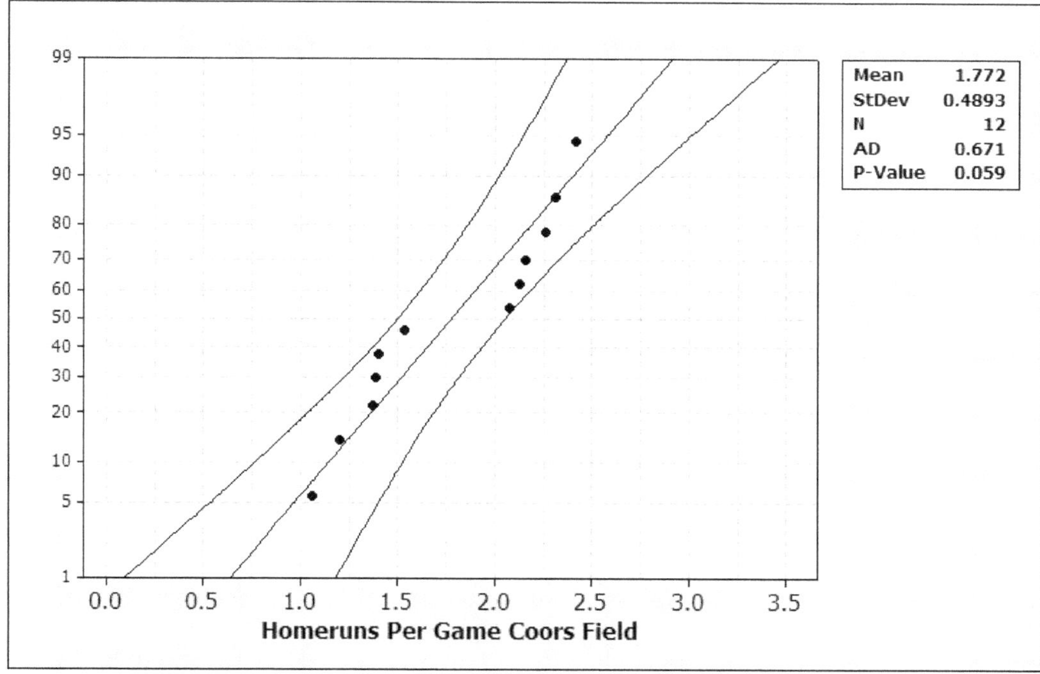

Step 3: Set Level of Significance

As previously stated, we will use 0.05 for all hypothesis tests in this book.

Step 4: Calculate Test Statistic

The sample mean, $\bar{X}$, is 1.77 and standard deviation, S, is 0.49. Substituting appropriately into the test statistic equation: $t_{stat} = \frac{\bar{X} - \mu_o}{\frac{S}{\sqrt{n}}} = \frac{1.77 - 1.55}{\frac{0.49}{\sqrt{12}}} = \frac{0.22}{0.141} = 1.56$

Step 5: Calculate p-value

We begin with the degrees of freedom, DF, equal to n – 1, which for this example comes to 11. Using the absolute value of our t_{stat} or 1.56, we search the T-table, a portion of which is provided in Figure 7.9, finding where 1.56 falls within the DF row of 11. From inspection, we find that the t_{stat} of 1.56 falls between 1.363 and 1.796, with right-tail probabilities of 0.10 and 0.05, respectively. Therefore, the right-tail probability for our t_{stat} of 1.56 must fall between 0.05 and 0.10. With the alternative hypothesis being "greater than," this results in the p-value for our test being between 0.05 and 0.10.

(If our alternative would have been "not equal," then we would have doubled these end points to say the p-value falls between 0.10 and 0.20.)

FIGURE 7.9: Portion of T-Table for Example 6.

T-Table: t Distribution Confidence Interval and Critical Values

	Confidence Level					
	80%	90%	95%	98%	99%	99.8%
	Right Tail Probability					
df	$t_{0.10}$	$t_{0.05}$	$t_{0.025}$	$t_{0.01}$	$t_{0.005}$	$t_{0.001}$
1	3.078	6.314	12.706	31.821	63.657	318.289
2	1.886	2.920	4.303	6.965	9.925	22.328
..	...	...	...	...	...	...
..	...	...	...	...	...	...
11	1.363	1.796	2.201	2.718	3.106	4.025

Step 6: Conclusion

With the p-value being between 0.05 and 0.10, we know that the p-value is greater than the 0.05 level of significance. Therefore, we fail to reject the null hypothesis. We would conclude, based on our sample data of home run records from 2000 through 2011, that there is not enough statistical evidence to conclude that Coors Field allows more home runs, on average, than the 1.55 home runs allowed by other stadiums. Annotated Minitab output is provided in Table 7.6, where again the p-value found by the software agrees with the hand calculation.

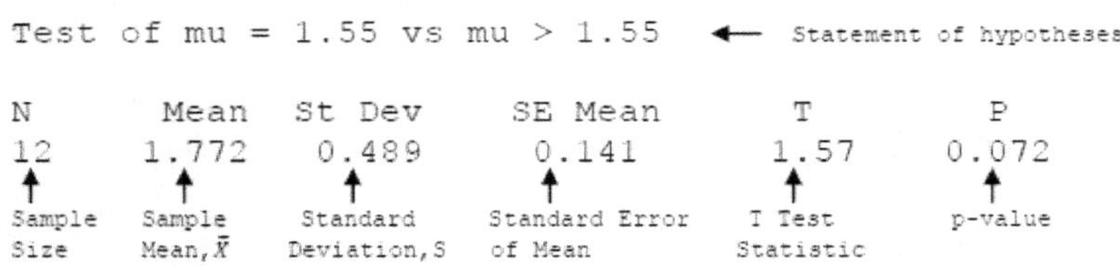

```
Test of mu = 1.55 vs mu > 1.55   ← Statement of hypotheses

N        Mean    St Dev     SE Mean         T         P
12       1.772   0.489      0.141           1.57      0.072
↑        ↑       ↑          ↑               ↑         ↑
Sample   Sample  Standard   Standard Error  T Test    p-value
Size     Mean,x̄  Deviation,S of Mean        Statistic
```

TABLE 7.6: Annotated Minitab output for Example 6—hypothesis test of one mean using 2000–2011 MLB data.

7.4 Errors and Power in Decision Making

As we are working with sample data—and therefore sample error is present—this leads to possible errors when making statistical decisions. We may reject the null hypothesis, when in reality this null hypothesis is true and should not have been rejected, or we may fail to reject the null hypothesis, when in reality this null hypothesis was false and should have been rejected. The first of these errors is referred to as a **Type I error** and the second is referred to as a **Type II error**. Figure 7.10 provides the four possible decision outcomes of a significance test.

FIGURE 7.10: The four possible decision outcomes when performing a significance test.

Statistical Decision	Reality	
	Ho is True	**Ho is False**
Reject Ho	Type I Error	Correct
Fail to Reject Ho	Correct	Type II Error

Consider, for example, Roger Clemens's perjury trial, which was concluded in June of 2012. Clemens was accused of lying to Congress about his use of steroids while pitching for the New York Yankees. Once the evidence was presented, a decision was made where he was acquitted of all charges related to lying to Congress about steroid use. The hypotheses set up was as follows:

Ho: Clemens did not lie to Congress.

Ha: Clemens did lie to Congress.

Prior to the final decision, the options were:

1. Fail to reject the null hypothesis—there was not enough evidence brought forward to conclude Clemens lied about his use of steroids; he did not commit perjury; or

2. Reject the null hypothesis—there was enough evidence brought forward to conclude Clemens lied about his use of steroids; he committed perjury.

The reality is either he did or did not lie to Congress, but only Clemens knows this for certain. This sets up the four possible outcomes outlined above in Figure 7.10:

- If Clemens was telling the truth, then a correct decision was made. This would be one where the null hypothesis was not rejected, and in reality the null hypothesis was true.

- If he was lying, then an incorrect decision was made. This would be one where the null hypothesis was not rejected, but in reality the null hypothesis was false. A Type II error was made.

- If Clemens was lying and he would have been convicted, then a correct decision would have been made. This would have been one where the null hypothesis was rejected, and in reality the null hypothesis was false.

- If he was telling the truth but would have been convicted, then an incorrect decision would have been made. This would be one where the null hypothesis was rejected, but in reality the null hypothesis was true. A Type I error would have been made.

Of course, the only way to know if an error was made is to know which hypothesis is true, and in reality we often don't know if our decision is correct. However, we can control the probabil ty in making an incorrect decision. To help protect against making a Type I error, we set the probability of this at an acceptably low level. This is our level of significance, alpha (α), we set at 0.05 in Step 3 of our hypothesis steps. For our tests, we have set as 0.05 the probability of incorrectly rejecting a null hypothesis that was true.

On the other hand, we would like to assign as high a probability as possible to correctly reject a false null hypothesis. In the Clemens example, this would be finding him guilty of perjury, when in fact he did perjure himself. We call this probability of correctly rejecting a false null hypothesis the **power of the test**. The power of a test is equal to one minus the probability of making a Type II error called beta (β):

$$\text{Power} = 1 - \beta$$

Considering all of these relationships, one might ask, "Why not set alpha so low as to render making a Type I error practically impossible?" The reason is simple: Imagine if alpha was set so low that Congress couldn't find Clemens guilty—essentially guaranteeing that he is found not guilty. This raises the probability of not finding him guilty if he truly is guilty. That is, they would increase the probability of making a Type II error, and in turn decrease the power of the test! Tying this together: If we decrease α, we increase β, and decrease power. The error probabilities have an inverse relationship, while α and power have a direct relationship.

7.5 Statistical Significance and Practical Significance

When we reach a statistically significant conclusion, we are saying that the sample data produced an unlikely result from that of the null hypothesis, and because of this we decide against the null hypothesis being true. This is called **statistical significance**. But this is just a statistical result; it does not mean the result has **practical significance**. Practical significance is exactly what it sounds like: Does the conclusion have any practical importance?

Consider the following: You are the trainer in charge of all student athletes at a large university, say Ohio State. In 2010, this encompassed 1002 student athletes across 31 sports (see www.fldcu.org). You are researching a new exercise designed to improve the vertical jumping ability of these athletes. After three months, you measure their jumps against their measurement prior to beginning the study. The results show an average improvement of a quarter inch (0.25 inches), with a standard deviation of one-tenth of an inch (0.1). When you conduct a t-test for this mean using the sample size of 656, you get a p-value of 0.000, indicating a statistically significant improvement from using this new exercise. However, do you really believe any coach is going to find this to be a practical improvement—that after three months, his or her athlete only gained an average of a quarter inch in their vertical jump? Very unlikely. The results may have been statistically significant, but they held no practical importance.

Expressions and Formulas

1. General format for null and alternative hypotheses—remember, select only ONE alternative.

One Proportion

$$H_o: p = p_o$$
$$H_a: p < p_o \text{ or } H_a: p > p_o \text{ or } H_a: p \neq p_o$$

One Mean

$$H_o: \mu = \mu_o$$
$$H_a: \mu < \mu_o \text{ or } H_a: \mu > \mu_o \text{ or } H_a: \mu \neq \mu_o$$

2. Test Statistic

One Proportion

$Z_{stat} = \dfrac{\hat{p} - p_o}{\sqrt{\dfrac{p_o(1 - p_o)}{n}}}$ where $\hat{p}$ is the sample proportion, p_o is the hypothesized proportion, and n is sample size.

One Mean

$t_{stat} = \dfrac{\bar{X} - \mu_o}{\dfrac{S}{\sqrt{n}}}$ where $\bar{X}$ is the sample mean, is the hypothesized mean, S is the sample standard deviation, and n is the sample size.

3. Power of a Significance Test

Power = 1 - β, the probability of correctly rejecting a null hypothesis (power) is equal to one minus the probability of making a Type II error (β).

Comparing Two Groups 8

In Chapters 6 and 7, we introduced ourselves to statistical inference, where our focus was on one population: either a proportion or a mean. But there are many circumstances where our interests involved more than one proportion. For example, we might want to compare:

1. Winning percentages for teams playing at home in college football and the NFL;
2. Third versus fourth round scores for players on the PGA tour;
3. NFL combine times in the 40-yard dash between the SEC and Big 10 conferences;
4. Salaries across the NFL, NBA, and MLB.

In each of these examples, we are comparing more than one population. However, as in those previous chapters, our consideration still involves either a proportion (Example 1) or mean (Examples 2, 3, and 4). To begin, we will focus only on the comparison of two groups (Examples 1, 2, and 3) and save more than two groups (Example 4) for the next chapter.

When we compare two groups, we use techniques associated with **bivariate** statistical methods. The term, bivariate, is defined exactly as it intuitively sounds: two variables. In such cases, we commonly have two variables: an outcome variable and a binary variable that specifies the two groups. In statistics, we associate the outcome variable as the **response variable** and the binary variable as the **explanatory variable**. The methods to analyze such data look to see how the response variable depends on, or is explained by, the groups of the explanatory variable. Based on previous chapters, you should be able to determine that the first example would involve a comparison of proportions and the remaining examples compare means. Looking at the first three examples above, we have the following:

1. A response variable, home winning percentages, depending on the explanatory variable of the games being college or professional.
2. A response variable, golf scores, depending on the explanatory variable of whether the score came in the third or fourth round.
3. A response variable, 40-yard dash times at NFL combine, depending on the explanatory variable of the player coming from the SEC or Big 10 Conference.

Take a closer look at Example 2, comparing PGA scores over the last two rounds of a tournament. If you were to conduct this analysis, would the data collection make more sense to compare the scores for these rounds using:

A. The same player (e.g., Tiger Woods's round 3 score to his round 4 score; Phil Mickelson's round 3 score to his round 4 score, etc.), or

B. Different players (e.g., compare Tiger's round 3 score to Phil's round 4 score, vice versa)?

Hopefully, you selected method "A," especially if your interest is in analyzing consistency. What sense would it make to compare Phil's third round score to Tiger's fourth round score? Maybe Phil shot 72 both rounds while Tiger shot 74 70. Phil's difference of zero shows consistency, while Tiger's difference of 4 does not. If you compared the rounds between players there would be no difference: in one round Phil was two shots better, and in the other round Phil was two shots worse. Situations where the data collection makes more sense to take two measurements from the same subject is referred to as **dependent samples** or a **matched pairs** design. In Example 3, where there are two distinct groups—SEC or Big 10—this is referred to as **independent samples**.

Regardless of proportions or means, independent or dependent, the focus is to determine if a statistical difference exists between the groups. The steps to hypothesis testing presented in Chapter 7 remain in play here in Chapter 8, as well as subsequent chapters; we just need to adjust our thoughts as we are now examining two populations. Therefore, we edit these steps to represent our current topic. Since a general practice is to test to see if there is a difference between the populations, the hypothesis statements are often stated using a null value of zero and a two-side "not equal" to zero alternative hypothesis .

Prior to our discussion on hypothesis testing for two groups, we must attend to a new concept when discussing the comparing of two groups, particularly with proportions and independent samples. This is the concept of equal variances across the two groups. If we can assume they are equal, then a **pooled estimate** is used in calculating the test statistic. If not, then an **unpooled estimate** is used. The use of pooled methods provides for a slightly more powerful test—lessening the chance of making a Type I error—however, serious error can result when the variances are assumed equal when they are not. Therefore, if you want to *assume* equal variances, you need to first check the reasonableness of this assumption.

8.1 Pooled versus Unpooled Variance Estimates

The easy one first. For testing two proportions, the use of pooled estimates rests simply with the null hypothesis, Ho. If the null hypothesis is testing "no difference," we have Ho: $p_1 = p_2$, then we calculate the test statistic based on the pooled estimate. In essence, since we are assuming the null hypothesis is true, an assumption of all hypothesis tests, we will "lump" the two proportions into one. However, in the event you run into a situation where the null hypothesis is something besides *no difference*, the proper test statistic can be found at the end of the chapter under *Expressions and Formulas*.

When testing two independent means, we focus on the variances of the two samples by comparing the two sample standard deviations. To illustrate the concept, we return to the Duke basketball example from Chapter 1. Figure 8.1 presents the height data for the two teams.

FIGURE 8.1: Statistics for men and women basketball teams at Duke University in 2011–2012.

Team	Mean	Variance	Standard Deviation
Men	78.46	14.90	3.86
Women	72.08	12.60	3.55

As Figure 8.1 illustrates, there is a large difference is mean heights—almost four and one half inches—between the two teams. Yet the variances, and subsequently the standard deviations, are quite similar. In such instances where the variances are reasonably close, one can combine this information into a **pooled standard deviation**.

RULE OF THUMB: A common practice in determining if the two variances are equal is to compare the two standard deviations. If the larger standard deviation is not more than **double** the smaller standard deviation, then we deem the variances equal and pool their information. Using the statistics in Figure 8.1 we would pool this information as the larger standard deviation, 3.86, is not more than double 3.55, the smaller standard deviation. Caution should be taken when using the pooled estimate if the sample sizes are not equal or not very similar. In the Duke basketball example, there were 13 males and 12 females. One could consider these sample sizes as similar. Unfortunately, there is no rule of thumb to define "similar."

8.2 Performing Hypothesis Tests of Two Groups

Step 1: State Hypotheses[1]

When considering the difference, unless the question specifically states the order, one does not have to be concerned with which population gets labeled as "1" and the other as "2." However, we need to be consistent with the order throughout the statistical process. For example, if we were comparing MLB attendance figures between the American and National leagues and we decided to label alphabetically, then we would need to consistently use American league data as group 1 and National league data as group 2. This becomes especially important in the decision making, so one correctly interprets any differences. The order must also be considered when applying software. Some software packages will go alphabetically, while others will calculate the difference based on which column of

1 In Chapter 7, our null value could take on any value. That is true here as well. Frequently, however, when we test two groups, our interest is in whether or not there is a difference. Therefore, the text will focus on such practices.

data is first entered. A practice one might consider is the use of specific subscripts which identify the groups. For instance, one might use the subscript "A" for American and "N" for National.

Two Proportions

$$H_o: p_1 - p_2 = 0$$

$$H_a: p_1 - p_2 \neq 0$$

Two Independent Means

$$H_o: \mu_1 - \mu_2 = 0$$

$$H_a: \mu_1 - \mu_2 \neq 0$$

Matched Pairs or Dependent Means

$$H_o: \mu_d = 0$$

$$H_a: \mu_d \neq 0$$

HINT: If you know in advance the sample proportions or the sample means, then if you arrange the hypotheses with the population parameter notation with the larger sample statistic first, you will always end up with a positive test statistic.

Step 2: Check Assumptions

Two Proportions

- Categorical response variable for two groups.
- Independent samples from two populations and large enough sample sizes to have at least 10 "successes" and 10 "failures" in each sample._

Two Independent Means

- Quantitative response variable for two groups.
- Independent samples from two populations that are approximately normal; and if not, then each sample is sufficiently large (each sample size at least 30). Recall the *Central Limit Theorem*.
- The two population variances are equal.

Matched Pairs or Dependent Means

- Quantitative response variable.
- Dependent samples where the differences are approximately normal; and if not, then the **number of differences** is sufficiently large (at least 30).

Step 3: Set Level of Significance

Recall we have set this at 0.05 or 5% for all significance tests in this book.

Step 4: Calculate Test Statistic

Two Proportions

$$Z_{stat} = \frac{(\hat{p}_1 - \hat{p}_2) - 0}{\sqrt{\hat{p}(1-\hat{p})\left(\frac{1}{n_1} + \frac{1}{n_2}\right)}}$$ where $\hat{p}$ in the denominator is found by:

$$\hat{p} = \frac{\textit{Total number of "successes" in both samples}}{\textit{Sum of both sample sizes}}$$

This "$\hat{p}$" is referred to as a **pooled estimate**, since it pools together the total number of successes and sample sizes from the two samples.

Two Independent Means

- If assumption of equal variances IS satisfied:

$$t_{stat} = \frac{(\bar{X}_1 - \bar{X}_2) - 0}{s_p\sqrt{\frac{1}{n_1} + \frac{1}{n_2}}}$$ where S_p is found by:

$$S_p = \sqrt{\frac{(n_1 - 1)S_1^2 + (n_2 - 1)S_2^2}{n_1 + n_2 - 2}}$$ This "S_p" is referred to as the **pooled estimate**, as it pools together the information from the two samples. The degrees of freedom, DF, when using the pooled estimate are found by DF = $n_1 + n_2 - 2$.

- If assumption of equal variances is NOT satisfied:

$$t_{stat} = \frac{(\bar{X}_1 - \bar{X}_2) - 0}{\sqrt{\frac{S_1^2}{n_1} + \frac{S_2^2}{n_2}}}$$

In either equation, $\bar{x}_1$, S_1, and n_1 represent, respectively, the mean, standard deviation, and sample size of one of the samples, and $\bar{x}_2$, S_2, and n_2 represent, respectively, the mean, standard deviation, and sample size of the second sample. The degrees of freedom, DF, when using the unpooled estimate do not provide any insight into the application of the method and are rather complicated. Therefore, the equation for finding the degrees of freedom when using the unpooled estimate are relegated to the *Expressions and Formulas* section located at the end of the chapter.

Matched Pairs or Dependent Means

$$t_{stat} = \frac{\bar{X}_d - 0}{\frac{S_d}{\sqrt{n}}}$$

where $\bar{X}_d$ and S_d represent, respectively, the mean and standard deviation of the paired differences, while n represents the number of paired differences in the sample. The degrees of freedom, DF, are found by DF = n - 1.

NOTE: In each test statistic equation we subtract zero. This is consistent with test statistics from Chapter 7 where we subtracted the null value, either p_0 or u_0. When comparing two groups we stated that the null value is typically zero; no difference between the two groups. Therefore, this p_0 or u_0 value is zero. However, if the problem requires a test of a difference besides zero, you would simply substitute in this number. For example, if you wanted to research whether current NBA players are more than two inches taller on average than NBA players from the 1960s, then u_0 would be 2 instead of 0.

Step 5: Calculate p-value

The method of finding the p-value will follow the same process as that in Chapter 7. However, many of these calculations, including the p-value, will be furnished by software. Our main focus will be on correctly interpreting this output.

Step 6: Conclusion

As we learned in Chapter 7, if the p-value is less than 0.05, we will reject our null hypothesis in favor of the alternative. By doing so, we will have decided there was enough statistical evidence to conclude a difference between the two populations. Again, this rejection of the null hypothesis would be considered statistically significant.

8.3 Confidence Intervals for Comparing Two Groups

Recall that in Chapter 6 we offered a general confidence interval formula. We restate that expression below:

Sample Statistic $\pm$ *Margin of Error*

where the

Margin of Error = Multiplier x Standard Error

Combining our notation for hypothesis testing, we can formulate confidence intervals for the comparison of two groups. The multipliers are found just as they were presented in Chapter 6.

Two Proportions

$$(\hat{p}_1 - \hat{p}_2) \pm Z^* \sqrt{\hat{p}(1 - \hat{p})\left(\frac{1}{n_1} + \frac{1}{n_2}\right)}$$

Two Independent Means

Equal variances assumed: $(\bar{X}_1 - \bar{X}_2) \pm t^* S_p \sqrt{\frac{1}{n_1} + \frac{1}{n_2}}$

Equal variances NOT assumed: $(\bar{X}_1 - \bar{X}_2) \pm t^* \sqrt{\frac{S_1^2}{n_1} + \frac{S_2^2}{n_2}}$

Matched Pairs or Dependent Means

$$X_d \pm t^* \frac{S_d}{\sqrt{n}}$$

When interpreting the intervals, remember that we are estimating a *difference* between two groups which must be reflected in our final analysis. If using confidence intervals in support of a significance test, then we simply inspect the interval to see if the value of zero is captured by the interval. If zero is contained within the interval, then we would not reject the null hypothesis. Contrarily, if the interval does not contain zero, then we would reject the null hypothesis.

8.4 Hypothesis Test and Confidence Interval: Two Proportions

Example 1: Home Field Advantage in Football: College versus the NFL

In Chapter 7, we determined that a home field advantage exists in both the collegiate and professional levels; statistically more than 50% of the time the home team won. The question now becomes, "Is there a statistically significant advantage to playing at home, depending on which level the games are played?" We can use the same data, except we apply methods for comparing two proportions. Resetting the 2011 data, we had:

College: Home team won 483 of 748 games for a sample proportion of 0.646.

NFL: Home team won 145 of 256 games for a sample proportion is 0.566.

Step 1: State Hypotheses

Using specific subscripts, we apply "C" to represent the college data, and "N" for the NFL data.

$$H_0: p_C - p_N = 0 \text{ versus } H_a: p_C - p_N \neq 0$$

Step 2: Check Assumptions

We have a categorical variable, win or lose, from two groups, college and professional football. These two groups are independent—one cannot be both a college and professional team (although many fans in the SEC believe their teams are better than some NFL teams!). To complete our assumptions, we have at least 10 "successes," the number of home wins, and at least 10 "failures," the number of home losses, in both the college and professional samples. The number of wins, 483 and 145, and number of losses, 265 and 111, satisfy the at least 10 requirements.

Step 3: Set Level of Significance

As previously stated, we will use 0.05 for all hypothesis tests in this book.

Step 4: Calculate Test Statistic

First, since our null hypothesis is assuming equal proportions, we must calculate the pooled estimate:

$$\hat{p} = \frac{\textit{Total number of "successes" in both samples}}{\textit{Sum of both sample sizes}} = \frac{483+145}{748+256} = \frac{628}{1004} = 0.625$$

Next, we calculate 0.646 as the sample proportion for college home games and 0.566 as the sample proportion for professional home games. With our hypotheses written as "college minus professional," we set up our test statistic:

$$Z_{stat} = \frac{(\hat{p}_C - \hat{p}_N) - 0}{\sqrt{\hat{p}(1-\hat{p})\left(\frac{1}{n_C}+\frac{1}{n_N}\right)}} = \frac{0.646 - 0.566}{\sqrt{0.625(1-0.625)\left(\frac{1}{748}+\frac{1}{256}\right)}} = \frac{0.08}{\sqrt{0.0012}} = \frac{0.08}{0.035} = 2.29$$

Step 5: Calculate p-value

With the alternative hypothesis being "not equal," we get the p-value by doubling the probability of getting greater than the $|Z_{stat}|$ or 2.29 in the Z-Table. From the Z-Table, a portion of which is provided in Figure 8.2, the cumulative probability for 2.29 is 0.9890, making the probability 0.011 of getting greater than 2.29. Finally, since the alternative is "not equal," we double this probability to arrive at the p-value of 0.022 for this test.

FIGURE 8.2: Portion of standard normal table for Example 1.

Z-Table (continued): Standard Normal Cumulative Probabilities

Cumulative probability (area to LEFT) of Positive Z-values

Z	0.00	0.01	0.02	0.03	0.04	0.05	0.06	0.07	0.08	0.09
0.0	0.5000	0.5040	0.5080	0.5120	0.5160	0.5199	0.5239	0.5279	0.5319	0.5359
0.1	0.5398	0.5438	0.5478	0.5517	0.5557	0.5596	0.5636	0.5675	0.5714	0.5753
0.2	0.5793	0.5832	0.5871	0.5910	0.5948	0.5987	0.6026	0.6064	0.6103	0.6141
...	...	...	...	...	...	...	...	...	...	...
...	...	...	...	...	...	...	...	...	...	...
2.2	0.9861	0.9864	0.9868	0.9871	0.9875	0.9878	0.9881	0.9884	0.9887	0.9890

Step 6: Conclusion

Comparing the p-value of 0.022 to 0.05, we find the p-value to be smaller. Therefore, we reject the null hypothesis in favor of the alternative. We would conclude, based on our sample data of college and NFL games played in 2011, that there is a statistically significant difference in home win percentage between the collegiate and professional levels.

In constructing a 95% confidence interval for this difference, we have:

$$(\hat{p}_C - \hat{p}_N) \pm Z^* \sqrt{\hat{p}(1 - \hat{p})\left(\frac{1}{n_C} + \frac{1}{n_N}\right)} = 0.08 \pm 1.96 * 0.035 = 0.08 \pm 0.069$$

This results in a 95% confidence interval for the difference in home field winning percentage between college and professional football is from 0.011 to 0.149 or from 1.1% to 14.9%. With the interval not including zero, the interval would support our conclusion that there is a difference in playing at home in college versus the NFL.

And this is where *how* this difference is calculated matters. Since we found the difference between the two groups by taking College *minus* NFL and this estimated difference is positive, we can further state that the difference appears to be in favor of the college teams. That is, there is statistical evidence to say that home teams in college win at a higher percentage than teams playing at home in the NFL. In Section 8.8, we will learn more about making these one-sided test decisions when our results are based on a two-sided alternative. Annotated Minitab output is provided in Table 8.1. The small differences in what we get by hand compared to software results - for instance the Z test statistics of 2.29 (hand) versus 2.26 (Minitab) - can be attributed to rounding in the hand calculations.

```
Sample     X     N   Sample p
College  483   748   0.645722 ←— sample data "group 1"
NFL      145   256   0.566406 ←— sample data "group 2"

Difference = p (1) - p (2) ←— difference is group 1 - group 2
Estimate for difference:   0.0793 ←— difference in sample proportions
95% CI for difference:   (0.0096, 0.1490) ←— 95% confidence interval
Test for difference = 0 (vs not = 0): ←— Alternative is "not equal"
Z = 2.26 P-Value = 0.024
      ↑                  ↑
Z test statistic      p-value
```

TABLE 8.1: Annotated Minitab output for Example 1—test and confidence interval of two proportions comparing home field winning percentage between college football and the NFL.

Example 2: Foul Shooting—NBA versus WNBA

Is there a difference in foul shooting between the two professional basketball leagues? Are men better/worse foul shooters than women? Using data from the 2011 season (the last full NBA season) as sample data, can one conclude there is a difference in foul shooting percentage between men and women? Detractors of women's basketball might argue that they use a smaller basketball compared to men, making shooting easier—the WNBA ball is about one inch smaller in circumference than an NBA ball (see www.wnba.com). However, proponents might counter that most likely a women's hand, on average, is smaller than a man's hand, making shooting a NBA-size basketball more difficult. Since the court dimensions are the same in both leagues, we will ignore this argument. In both leagues the foul line is set 15 feet from the front of the backboard and the top of the rim is 10 feet from the floor. Based on team statistics for the 2011 WNBA and 2010–2011 NBA seasons, we have (courtesy www.nba.com):

WNBA: 5860 foul shots made out of 7613 attempted, a percentage of 0.770.

NBA: 45,760 foul shots made out of 59,937 attempted, a percentage of 0.763.

Step 1: State Hypotheses

Using specific subscripts, we apply "W" to represent the women's data, and "M" for the men's data.

$$H_o: p_W - p_M = 0 \text{ versus } H_a: p_W - p_M \neq 0$$

Step 2: Check Assumptions

We have a categorical variable, make shot or miss shot, from two groups, women's and men's professional basketball. These two groups are independent—one cannot be a player in both leagues. To complete our assumptions, we have at least 10 "successes," made foul shots, and at least 10 "failures," missed foul shots, in both the women's and men's samples. This assumption is clearly satisfied.

Step 3: Set Level of Significance

As previously stated, we will use 0.05 for all hypothesis tests in this book.

Step 4: Calculate Test Statistic

First, since we our null hypothesis is assuming equal proportions, we must calculate the pooled estimate:

$$\hat{p} = \frac{\text{Total number of "successes" in both samples}}{\text{Sum of both sample sizes}} = \frac{5860 + 45760}{7613 + 59937} = \frac{51620}{67550} = 0.764$$

Next, we use 0.770 as the sample proportion for women's foul shooting and 0.763 as the sample proportion for men's foul shooting. With our hypotheses written as "women minus men," we set up our test statistic:

$$Z_{stat} = \frac{(\hat{p}_W - \hat{p}_M) - 0}{\sqrt{\hat{p}(1-\hat{p})\left(\frac{1}{n_W} + \frac{1}{n_M}\right)}} = \frac{0.770 - 0.763}{\sqrt{0.764(1-0.764)\left(\frac{1}{7613} + \frac{1}{59937}\right)}} = \frac{0.007}{\sqrt{0.000027}} = \frac{0.007}{0.0052} = 1.35$$

Step 5: Calculate p-value

With the alternative hypothesis being "not equal," we get the p-value by doubling the probability of getting greater than the $|Z_{stat}|$ or 1.35 in the Z-Table. From the Z-Table, a portion of which is provided in Figure 8.3, the cumulative probability for 1.35 is 0.9115, making the probability 0.0885 of getting greater than 1.35. Finally, since the alternative is "not equal," we double this probability to arrive at the p-value of 0.177 for this test.

FIGURE 8.3: Portion of standard normal table for Example 2.

Z-Table (continued): Standard Normal Cumulative Probabilities

Cumulative probability (area to LEFT) of Positive Z-values

Z	0.00	0.01	0.02	0.03	0.04	0.05	0.06	0.07	0.08	0.09
0.0	0.5000	0.5040	0.5080	0.5120	0.5160	0.5199	0.5239	0.5279	0.5319	0.5359
0.1	0.5398	0.5438	0.5478	0.5517	0.5557	0.5596	0.5636	0.5675	0.5714	0.5753
0.2	0.5793	0.5832	0.5871	0.5910	0.5948	0.5987	0.6026	0.6064	0.6103	0.6141
...	...	...	...	...	...	...	...	...	...	...
...	...	...	...	...	...	...	...	...	...	...
1.3	0.9032	0.9049	0.9066	0.9082	0.9099	0.9115	0.9131	0.9147	0.9162	0.9177

Step 6: Conclusion

Comparing the p-value of 0.177 to 0.05, we find the p-value to be greater. Therefore, we fail to reject the null hypothesis. We would conclude, based on our sample data of WNBA and NBA 2011 season statistics, that there is insufficient statistical evidence to conclude a difference in foul shooting percentage between women and men professional basketball players.

In constructing a 95% confidence interval for this difference, we have:

$$(\hat{p}_W - \hat{p}_M) \pm Z^* \sqrt{\hat{p}(1-\hat{p})\left(\frac{1}{n_W}+\frac{1}{n_M}\right)} = 0.007 \pm 1.96 * 0.0052 = 0.007 \pm 0.010$$

This results in a 95% confidence interval for the difference in home field winning percentage between college and professional football is from *negative* 0.003 to 0.017 or from *negative* 0.3% to 1.7%. Two things we should notice about this interval. One, with the interval including zero, the interval would support our conclusion that there is not a difference in foul shooting percentage between the two professional leagues. Two, we cannot have a negative proportion or negative percentage. This negative is a mathematical result but not a practical one. We use the negative to help define our answer; however, in practice, we would write this interval constraint as 0%. Annotated Minitab output is provided in Table 8.2.

```
Sample      X      N   Sample p
WNBA     5860   7613   0.769736 ←— sample data "group 1"
NBA     45760  59937   0.763468 ←— sample data "group 2"

Difference = p (1) - p (2) ←— difference is group 1 - group 2
Estimate for difference:  0.0063 ←— difference in sample proportions
95% CI for difference:  (-0.0038, 0.0163) ←— 95% confidence interval
Test for difference = 0 (vs not = 0): ←— Alternative is "not equal"
Z = 1.21  P-Value = 0.225
      ↑                   ↑
Z test statistic        p-value
```

TABLE 8.2: Annotated Minitab output for Example 2—test and confidence interval of two proportions comparing foul shooting percentages between WNBA and NBA.

The output provides a good opportunity to see the effect rounding can have on these calculations. In our hand calculations, we rounded the sample proportions to three decimal places, resulting in a 0.007 difference between the two statistics with a 1.35 test statistic. From the Minitab output in Table 8.2, this difference is 0.0063, resulting in a 1.21 test statistic. This also affects the p-values. If we had used the 0.0063 difference in our hand calculations the test statistic would have been 1.21, matching that of the software. Now imagine if the p-values were closer to 0.05, possibly producing conflicting conclusions; one method leading us to "reject the null hypothesis" and the other to "fail to reject the null hypothesis." When conducting hypothesis tests, one must be careful when making decisions where conclusions are influenced by rounding.

8.5 Hypothesis Test and Confidence Interval: Two Independent Means

Example 3: Difference in Home Runs between American and National Leagues

Since the advent of the designated hitter in the American League (AL), some debate has taken place on this making the AL more of a power-hitting league compared to the National League (NL). In order to test this theory, we use each team's home run total from the 2011 season as a sample of all years since 1973 when the rule was adopted. When games are played involving teams from both leagues—called interleague games—the game is played under the league rules of the home team (e.g., if played in the AL, then a designated hitter is used). The data from 2011 (from www.mlb.com) will reflect only games played *within* the league: AL versus AL and NL versus NL. From the data, we have the following summary measures:

$$\text{AL: } \bar{X} = 144.0 \text{ and } S = 34.2 \; n = 14$$

$$\text{NL: } \bar{X} = 129.6 \text{ and } S = 26.9 \; n = 16$$

Step 1: State Hypotheses

Using specific subscripts, we apply "A" to represent American League, and "N" to represent National League. Since the average home runs for the AL are more than the average homeruns for the NL, we will arrange our hypothesis to produce a positive test statistic.

$$H_o: \mu_A - \mu_N = 0 \text{ versus } H_a: \mu_A - \mu_N \neq 0$$

Step 2: Check Assumptions

We have a quantitative variable, home runs, for two groups: American and National Leagues. These two samples are independent—a team cannot play in both leagues at the same time. With a sample of 14 AL teams and 16 NL teams, we do not have a sufficiently large enough sample to assume the data is approximately normal; we must include **probability plots** to support this assumption. Recall from Chapter 7 that when using a probability plot, we are interested in a p-value *greater* than 0.05 in order to support our normal assumption. From Figure 8.4, we see that the p-values for the two groups exceed 0.05, supporting our assumption. The final assumption is that of equal variances for the two groups. Using the rule of thumb where the larger sample standard deviation is not more than twice the smaller sample standard deviation, our sample data produces sample standard deviations of 34.2 and 26.9, thus allowing us to assume the variances are equal. The difference in sample sizes, 14 and 16, we will consider to be similar enough for this example.

FIGURE 8.4: Probability plots for Example 3 comparing home run totals for American and National league teams during the 2011 season. With the p-values for each group exceeding 0.05, we assume the population of each group is approximately normal.

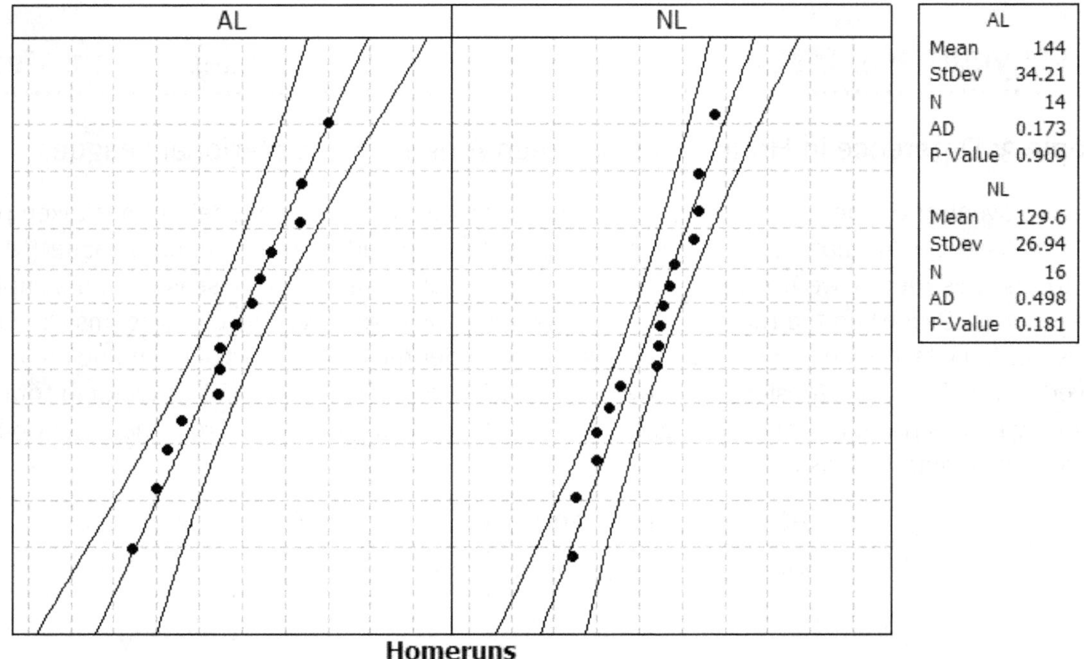

Step 3: Set Level of Significance

As previously stated, we will use 0.05 for all hypothesis tests in this book.

Step 4: Calculate Test Statistic

Since we are assuming equal variances, we first calculate the pooled estimate:[2]

$$S_p = \sqrt{\frac{(n_A - 1)S_A^2 + (n_N - 1)S_N^2}{n_A + n_N - 2}} = \sqrt{\frac{(14 - 1)34.2^2 + (16 - 1)26.9^2}{14 + 16 - 2}} = \sqrt{\frac{26059.47}{28}} = 30.51$$

Using this pooled estimate, we move forward to calculate the test statistic:

$$t_{stat} = \frac{(\bar{X}_A - \bar{X}_N) - 0}{S_p\sqrt{\frac{1}{n_A} + \frac{1}{n_N}}} = \frac{(144.0 - 129.6)}{30.51\sqrt{\frac{1}{14} + \frac{1}{16}}} = \frac{14.4}{11.17} = 1.29$$

2 In cases where the sample sizes are equal, this pooled standard deviation will just be the square root of the average of the two **sample variances**.

Step 5: Calculate p-value

Since we assumed equal variances, the calculation of the degrees of freedom are straightforward. The DF is found by $n_1 + n_2 - 2$, which comes to 28. Using the absolute value of our t_{stat} or 1.29, we search the T-Table, a portion of which is provided in Figure 8.5, finding where 1.29 falls within the DF row of 28. Within that row, we find that the first t-statistic we come to is 1.313, which corresponds to a right-tail probability of 0.10. That is, with a DF of 28, there is a 0.10 probability of getting a t-statistic of 1.313 or greater. With a t_{stat} of 1.29 preceding 1.303, there is more than a 0.100 probability of getting a t-statistic of 1.29 or greater. With the alternative hypothesis being "not equal," we must double this right-tail probability. This results in the p-value for our test being greater than 0.20.

FIGURE 8.5: Portion of T-Table for Example 3.

T-Table: t Distribution Confidence Interval and Critical Values

	Confidence Level					
	80%	90%	95%	98%	99%	99.8%
	Right Tail Probability					
df	$t_{0.10}$	$t_{0.05}$	$t_{0.025}$	$t_{0.01}$	$t_{0.005}$	$t_{0.001}$
1	3.078	6.314	12.706	31.821	63.657	318.289
2	1.886	2.920	4.303	6.965	9.925	22.328
..	...	...	...	...	...	...
..	...	...	...	...	...	...
28	1.313	1.701	2.048	2.467	2.763	3.408

Step 6: Conclusion

Comparing the p-value of 0.20 to 0.05, we find the p-value to be *greater*. Therefore, we *fail to reject* the null hypothesis. We would conclude, based on our sample data of home runs hit during the 2011 MLB season, that there is a lack of statistical evidence to conclude that, on average, American League teams hit more home runs than National League teams.

In constructing a 95% confidence interval for this difference, we have:

$$(\bar{X}_A - \bar{X}_N) \pm t * S_p\sqrt{\frac{1}{n_A} + \frac{1}{n_N}} = (144.0 - 129.6) \pm 2.048 * 30.51\sqrt{\frac{1}{14} + \frac{1}{16}} = 14.4 \pm 22.9$$

This results in a 95% confidence interval for the difference in mean home runs between American and National league teams is from -8.5 to 37.3. With the interval including zero, the interval would support our conclusion that there is no difference in mean home runs hit between the two leagues. Annotated Minitab output is provided in Table 8.3. As we can see, the hand calculations and those provided by Minitab are quite similar. The Minitab p-value of 0.207 agrees with our calculation that the p-value exceed 0.20.

```
League   N    Mean   StDev  SE Mean
AL      14   144.0    34.2      9.1 ←— sample data "group 1"
NL      16   129.6    26.9      6.7 ←— sample data "group 2"

Difference = mu (AL) - mu (NL) ←— difference is group 1 - group 2
Estimate for difference:  14.4 ←— difference in sample means
95% CI for difference: (-8.4, 37.3) ←— 95% confidence interval
T-Test for difference = 0 (vs not =): ←— Alternative is "not equal"
T-Value = 1.29 P-Value = 0.207  DF = 28
            ↑                ↑        ↑
     T test statistic     p-value  Degrees Freedom
Both use Pooled StDev = 30.5275 ←— Pooled estimate Sp
```

TABLE 8.3: Annotated Minitab output for Example 3—comparing mean home runs hit between American and National leagues.

Example 4: Difference in Salary between Professional Baseball and Football Players

One would be hard pressed to argue that professional athletes are suffering financially, especially in the major sports of baseball, basketball, and football. Of course, some of these sports come with injury risk shortening a player's career, making the high salary a potentially short-term gain. However, a question one may ask is if there is a difference between how much players are paid. For instance, is there a difference in average salary between Major League Baseball and professional football players? To answer this question, we take a random sample of 30 players from each of the two leagues with their salaries from the 2011 season (courtesy www.cbssports.com). The sample produced the following statistics in millions of dollars:

Baseball: = 3.77 and S = 4.86

Football: = 1.55 and S = 1.49

Step 1: State Hypotheses

Using specific subscripts, we apply "B" to represent baseball, and "F" to represent football. Since the average sample salary of baseball players is more than the average sample salary of football players, we will arrange our hypothesis to produce a positive test statistic.

$$H_o: \mu_B - \mu_F = 0 \text{ versus } H_a: \mu_B - \mu_F \neq 0$$

Step 2: Check Assumptions

We have a quantitative variable, salary, for two groups: players in MLB and the NFL. These two samples are independent—no player competed in both leagues. With a sample of 30 players, we have a sufficiently large enough sample to assume the data is approximately normal. Considering equal variances would not be appropriate in this case, since the sample standard deviation for baseball players (4.86) is more than double that for the football players (1.49). As a result, we will NOT use the pooled estimate.

Step 3: Set Level of Significance

As previously stated, we will use 0.05 for all hypothesis tests in this book.

Step 4: Calculate Test Statistic

Since we are cannot assume equal variances, our test statistic is:

$$t_{stat} = \frac{(\bar{X}_B - \bar{X}_F) - 0}{\sqrt{\frac{S_B^2}{n_B} + \frac{S_F^2}{n_F}}} = \frac{(3.77 - 1.55)}{\sqrt{\frac{4.86^2}{30} + \frac{1.49^2}{30}}} = \frac{2.22}{\sqrt{0.861}} = 2.39$$

Step 5: Calculate p-value

Since we are not assuming equal variances, making the calculation of the degrees of freedom messy, we will instead use software to guide us. Doing so results in 34 degrees of freedom. Using the absolute value of our t_{stat} or 2.39, we search the T-Table, a portion of which is provided in Figure 8.6, finding where 2.39 falls within the DF row of 30, since this is the closest degree of freedom without exceeding 34. Within that row, we find that 2.39 falls between t-statistics of 2.042 and 2.457, which correspond to a right-tail probability of 0.025 and 0.010, respectively. That is, with a DF of 30, there is a 0.010 to 0.025 probability of getting a t-statistic within the values of 2.042 and 2.457. With our t_{stat} of 2.39 falling within these two t-values, we can state that the probability of getting a t-statistic of 2.39 or greater is somewhere between 0.010 and 0.025. With the alternative hypothesis being "not equal," we must double this probability range. This results in the p-value for our test being greater than 0.020 but less than 0.050.

FIGURE 8.6: Portion of T-Table for Example 4.

T-Table: t Distribution Confidence Interval and Critical Values

	Confidence Level					
	80%	90%	95%	98%	99%	99.8%
	Right Tail Probability					
df	$t_{0.10}$	$t_{0.05}$	$t_{0.025}$	$t_{0.01}$	$t_{0.005}$	$t_{0.001}$
1	3.078	6.314	12.706	31.821	63.657	318.289
2	1.886	2.920	4.303	6.965	9.925	22.328
..	...	...	...	...	...	...
..	...	...	...	...	...	...
30	1.310	1.697	2.042	2.457	2.750	3.385

Step 6: Conclusion

Comparing the range of our p-value to 0.05, we find the p-value to be smaller. Even though we cannot get a more exact value from the T-Table, we know the range that it will fall is between 0.020 to 0.050. With *all* values in this range falling *below* 0.05, we reject the null hypothesis in favor of the alternative. We would conclude, based on our sample data of salaries for 30 major league baseball and football players, that there is a statistically significant difference in mean salary between these two professional leagues.

In constructing a 95% confidence interval for this difference, we have:

$$(\bar{X}_B - \bar{X}_F) \pm t * \sqrt{\frac{S_B^2}{n_B} + \frac{S_F^2}{n_F}} = (3.77 - 1.55) \pm 2.042 * \sqrt{\frac{4.86^2}{30} + \frac{1.49^2}{30}} = 2.22 \pm 1.89$$

This results in a 95% confidence interval for the difference in mean salary between professional baseball and football players is from 0.33 to 4.11 million dollars. With the interval not including zero, the interval would support our conclusion that there is a difference in mean salary between players in these two professional leagues.

Once again, *how* this difference is calculated matters. Since we found the difference between the two groups by taking Baseball *minus* Football and this estimated difference is positive, we can further state that on average, Major League Baseball players make more than those in the NFL. That is, there is statistical evidence to say that, on average, Major League Baseball players have higher salaries than those playing in the NFL. Annotated Minitab output is provided in Table 8.4. As it should, the p-value in the output, P-Value = 0.023, falls within our hand-calculated range of 0.020 to 0.050.

```
       N  Mean  StDev  SE Mean
MLB   30  3.77   4.86     0.89 ◄── sample data "group 1"
NFL   30  1.55   1.49     0.27 ◄── sample data "group 2"

Difference = mu (MLB) - mu (NFL) ◄── difference is group 1 - group 2
Estimate for difference: 2.211 ◄── difference in sample means
95% CI for difference: (0.324, 4.097) ◄── 95% confidence interval
T-Test of difference = 0 (vs not =): ◄── Alternative is "not equal"
T-Value = 2.38 P-Value = 0.023  DF = 34
           ▲                 ▲        ▲
   T test statistic       p-value  Degrees Freedom
```

TABLE 8.4: Annotated Minitab output for Example 4—comparing mean salaries for MLB and NFL players.

Example 5: Is There a Difference in Speed between SEC and Big10 Football Players?

Many pundits across the sports landscape—Lee Corso and Mark May of ESPN to name just two—have on multiple occasions remarked that the SEC has more speed than the Big10. This may or may not be true, and is certainly a hypothesis that offers a difficult solution. Measuring individual speed and combining that with overall team speed, only to take into consideration "playing" speed, registers on the near-impossible scale. However, one attempt could be made by using 40-yard dash speeds from the NFL combine. This metric might not matter all too much on the field, but it is used to measure one's speed. The combine itself provides some caution, since not all draft-eligible players attend the event. Some remain on their respective campuses awaiting a "Pro Day," when various NFL team associates swarm the colleges to hold individual tryouts. Running on familiar turf could influence the results, i.e., a player may run faster on his home field. Given this, using combine data will at least put all participants on common ground. Using data found at www.footballsfuture.com, we attempt to compare speed between the SEC and Big10 conferences. Player data is from the NFL combine years 2009 through 2012. The positions include running back, wide receivers, and secondary. For this particular example, we will start with the non-annotated Minitab output in Table 8.5, and use this output to research our hypotheses.

```
Conference   N    Mean   StDev  SE Mean
Big10       49  4.5596  0.0970    0.014
SEC         79  4.5415  0.0982    0.011

Difference = mu (Big10) - mu (SEC)
Estimate for difference:  0.0181
95% CI for difference:  (-0.0171, 0.0533)
T-Test of difference = 0 (vs not =):
T-Value = 1.02  P-Value = 0.311  DF = 126
Both use Pooled StDev = 0.0978
```

TABLE 8.5: Non-annotated Minitab output for Example 5—comparing mean forty times at NFL combine for Big10 and SEC players.

Using Table 8.5. we can run through all our hypothesis steps, including finding a confidence interval.

Step 1: State Hypotheses

Putting together the pieces "Difference = mu (Big10) - mu (SEC)" and "T-Test of difference = 0 (vs. not =)," we can formulate the hypotheses as:

$$H_o: \mu_B - \mu_S = 0 \text{ versus } H_a: \mu_B - \mu_S \ {}^{1}\ 0$$

where "B" represents the Big10 and "S" represents the SEC.

Step 2: Check Assumptions

Forty-yard times are quantitative and are sampled from two groups: Big 10 or SEC. These two groups are independent, as you can only be enrolled in one school at a time. The sample sizes are 49 for the Big10 and 79 for the SEC, found under the column labeled "N." Since both sample sizes are at least 30, we assume approximate normal distribution of our data. From the output, we have sample standard deviations of 0.0982 and 0.0970, with the larger not being twice the smaller. Therefore, pooled estimates can be used. We can see in the output that this was done where we find "Both use Pooled StDev = 0.0978." Although the sample sizes are quite different, we will assume equal variances based on the rule of thumb. However, students should be cautious of this as their instructor may suggest on NOT assuming equal variances whenever the sample sizes differ.

Step 3: Set Level of Significance

Per the book, we set our level of significance at 0.05.

Step 4: Calculate Test Statistic

Since pooled estimates were used, we have a pooled test statistic of 1.02, found in the output at "T-Value = 1.02."

Step 5: Calculate p-value

From the output, we have a p-value of 0.311 based on 126 degrees of freedom. These are easily recognized in the output by the "P-Value" and "DF" titles.

Step 6: Conclusion

Comparing p-value of 0.311 to 0.05, we find the p-value is larger. We cannot reject the null hypothesis. There is not enough statistical evidence to conclude that there is a difference in mean forty-yard time at the NFL combine between Big10 and SEC players. Since we are using this metric to measure speed, we can conclude that we do not have enough evidence to state that, on average, SEC players are faster than Big 10 players.

The 95% confidence interval supports our test findings. The reported confidence interval in Table 8.5 for estimated mean difference in forty-yard times is (-0.0171, 0.0533). This is located under the phrase "95% CI for difference." As this interval includes zero, we have further evidence that we cannot reject the null hypothesis.

8.6 Hypothesis Test and Confidence Interval: Matched Pairs

Example 6: In the NHL, Is There a Difference in How a Game Is Refereed between Regular and Post Seasons?

Hockey is a violent sport, where collisions are commonplace. Sometimes these actions result in penalties, which can have a negative effect—the penalized team being a man down—or a positive effect—the penalized team, if they can keep the opposition from scoring—may gain momentum. Players and fans of any sport should want, and expect, consistency in how games are refereed. There should not be a difference, for example, in the average penalties per game for a team between regular and post seasons. Using the 2010–2011 NHL season as a sample, we research this hypothesis to study if a difference does exist (see www.nhl.com). We use a paired data method, as comparing a team's performance from regular to post season makes more sense than comparing one team's regular season to another team's post season. We would, therefore, take two *measurements* of both teams. With 16 teams qualifying for the NHL play-offs, this leaves us with 16 sample differences: a team's per game penalties in the regular season versus post season.

From the data, where the difference, d, is found by taking Regular minus Post we have:

$\bar{X}_d = 0.84$ and $S_d = 1.23$

Step 1: State Hypotheses

$$H_o: \mu_d = 0 \text{ versus } H_a: \mu_d \neq 0$$

Step 2: Check Assumptions

We have a quantitative variable, penalties per game, for dependent samples: regular and post season statistics for the same 16 teams. With a sample of 16 teams, we have only 16 differences. This is not sufficiently large, so we will use a probability plot of the differences. From Figure 8.7, we can assume that the differences do follow a normal distribution as the p-value of 0.087 is greater than 0.05.

FIGURE 8.7: Probability plots for Example 6—regular and post season per game penalties in NHL. With the p-values exceeding 0.05, we assume the population is approximately normal.

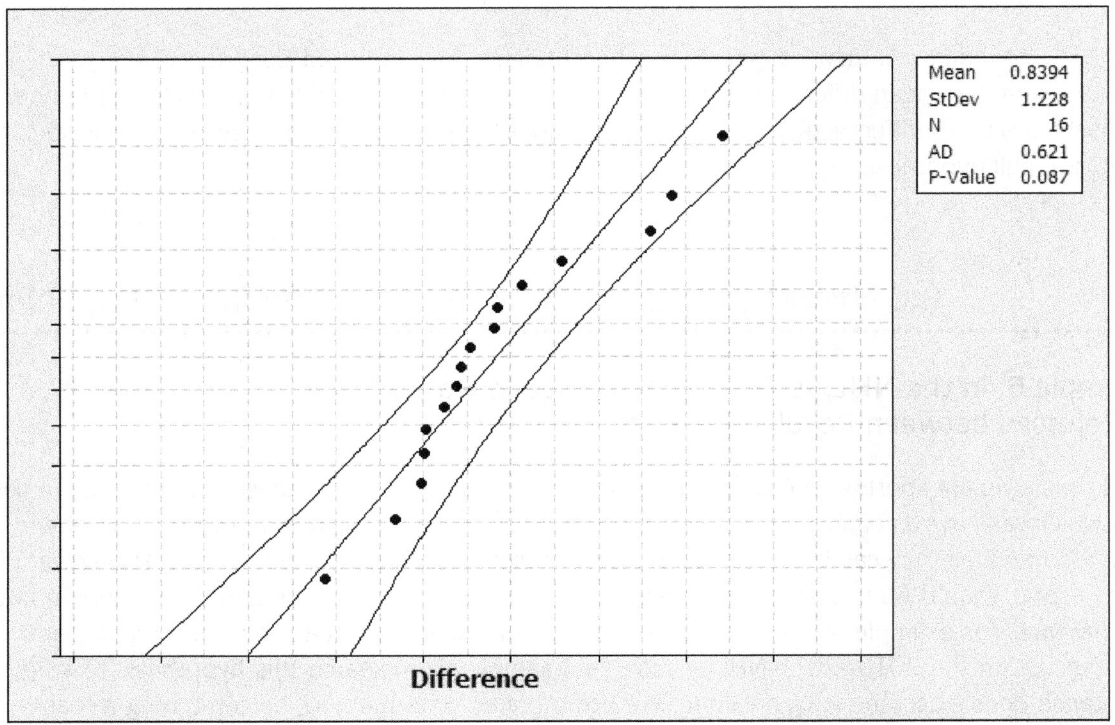

Step 3: Set Level of Significance

As previously stated, we will use 0.05 for all hypothesis tests in this book.

Step 4: Calculate Test Statistic

$$t_{stat} = \frac{\bar{x}_d - 0}{\frac{S_d}{\sqrt{n}}} = \frac{0.84}{\frac{1.23}{\sqrt{16}}} = \frac{0.84}{0.3075} = 2.73$$

Step 5: Calculate p-value

We begin with the degrees of freedom, DF, equal to n – 1, where n is the number of differences, 16. This makes our DF to 15. Using the absolute value of our t_{stat} or 2.73, we search the T-Table, a portion of which is provided in Figure 8.8, finding where 2.73 falls within the DF row of 15. From inspection, we find that the t_{stat} of 2.73 falls between 2.602 and 2.947, with right-tail probabilities of 0.010 and 0.005, respectively. Therefore, the right-tail probability for our t_{stat} of 2.73 must fall between 0.005

and 0.010. With the alternative hypothesis being "not equal," we must double this probability range. This results in the p-value for our test being greater than 0.010 but less than 0.020.

FIGURE 8.8: Portion of T-Table for Example 6.

T-Table: t Distribution Confidence Interval and Critical Values

	Confidence Level					
	80%	90%	95%	98%	99%	99.8%
	Right Tail Probability					
df	$t_{0.10}$	$t_{0.05}$	$t_{0.025}$	$t_{0.01}$	$t_{0.005}$	$t_{0.001}$
1	3.078	6.314	12.706	31.821	63.657	318.289
2	1.886	2.920	4.303	6.965	9.925	22.328
..	...	...	...	...	...	...
..	...	...	...	...	...	...
15	1.341	1.753	2.131	2.602	2.947	3.733

Step 6: Conclusion

Comparing the range of our p-value to 0.05, we find the p-value to be smaller. Even though we cannot get a more exact value, we know the range that it will fall is between 0.010 to 0.020. With *all* values in this range falling below 0.05, we reject the null hypothesis in favor of the alternative. We would conclude, based on our sample data of the 16 NHL play-off teams in the 2010–2011 season, that there is a statistically significant difference in mean number of penalties called per game between regular and post seasons.

In constructing a 95% confidence interval for this difference, we have:

$$\bar{X}_d \pm t * \frac{S_d}{\sqrt{n}} = 0.84 \pm 2.131 * \frac{1.23}{\sqrt{16}} = 0.84 \pm 0.66$$

This results in a 95% confidence interval for the mean difference per game penalties during the regular and post seasons of the NHL is from 0.18 to 1.50 penalties per game. With the interval not including zero, the interval would support our conclusion that there is a difference in per game penalties.

Again, *how* this difference is calculated matters. Since we found the difference between the two groups by taking Regular *minus* Post and this estimated difference is positive, we can further state that on average, there are more penalties called per game during the regular season compared to the post season in the NHL. Annotated Minitab output is provided in Table 8.6. As it should, the p-value in the output, P-Value = 0.015, falls within our hand-calculated range of 0.010 to 0.020.

```
              N    Mean   StDev   SE Mean
Regular      16   4.431   0.462     0.116 <-- sample data "group 1"
Post         16   3.592   1.069     0.267 <-- sample data "group 2"
Difference   16   0.839   1.228     0.307 <-- sample data for differences

95% CI for mean difference: (0.185, 1.494) <-- 95% confidence interval
T-Test of mean difference = 0 (vs not = 0): <-- Alternative is "not equal"
T-Value = 2.73 P-Value = 0.015
            ^                 ^
     T test statistic      p-value
```

TABLE 8.6: Annotated Minitab output for Example 6 —comparing penalties per game during regular and post season in NHL.

Example 7: Are PGA Tour Winners Consistent in Their Play over the Weekend?

The typical PGA tournament consists of four rounds, one played each day from Thursday through Sunday. Following the second round on Friday, a cut is made, where only the golfers meeting a specified criteria—for example, being in the top 70 or within 10 strokes of the leader—continue playing into the weekend. A question one might ask is: Are the tournament winners consistent in their weekend play? That is, is there evidence to indicate a disparity in one round compared to the other? Using as a sample the third and fourth round scores for winners during the 2011 PGA season, we have 46 events.[3] From the data (see www.espn.go.com/golf), where the difference, d, is found by taking Third Round minus Fourth Round, we have:

$$\bar{X}_d = -0.196 \text{ and } S_d = 3.48$$

Step 1: State Hypotheses

$$H_o: \mu_d = 0 \text{ versus } H_a: \mu_d \ ^1\ 0$$

Step 2: Check Assumptions

We have a quantitative variable, the number of strokes, for dependent samples: third and fourth round for the same 46 players. With a sample of 46 players, we have 46 differences. With the sample size being at least 30, we assume an approximately normal distribution.

Step 3: Set Level of Significance

As previously stated, we will use 0.05 for all hypothesis tests in this book.

---

3 There were 48 tournaments but the Barclays was stopped after three rounds due to weather, and one is the World Match Play, which does not follow a stroke format. Also, the Bob Hope Classic is five rounds, so the winner's fourth and fifth rounds were used.

Step 4: Calculate Test Statistic

$$t_{stat} = \frac{\bar{X}_d - 0}{\frac{S_d}{\sqrt{n}}} = \frac{-0.196}{\frac{3.48}{\sqrt{46}}} = \frac{-0.196}{0.513} = -0.382$$

Step 5: Calculate p-value

We begin with the degrees of freedom, DF, equal to n - 1, where n is the number of differences, 46. This makes our DF to 45, and we will use 40 from our T-Table since there is not a 45. Using the absolute value of our t_{stat} or 0.382, we search the T-Table, a portion of which is provided in Figure 8.9, finding where 0.382 falls within the DF row of 40. Starting in that row, the first t-statistic we come to is 1.303, which corresponds to a right-tail probability of 0.100. That is, with a DF of 40, there is a 0.100 probability of getting a t-statistic of 1.303 or greater. With our t_{stat} of 0.382 preceding 1.303, there is more than a 0.100 probability of getting a t-statistic of 0.382 or greater. With the alternative hypothesis being "not equal," we must double this right-tail probability. This results in the p-value for our test being greater than 0.200.

FIGURE 8.9: Portion of T-Table for Example 7.

T-Table: t Distribution Confidence Interval and Critical Values

	Confidence Level					
	80%	90%	95%	98%	99%	99.8%
	Right Tail Probability					
df	$t_{0.10}$	$t_{0.05}$	$t_{0.025}$	$t_{0.01}$	$t_{0.005}$	$t_{0.001}$
1	3.078	6.314	12.706	31.821	63.657	318.289
2	1.886	2.920	4.303	6.965	9.925	22.328
..	...	...	...	...	...	...
..	...	...	...	...	...	...
40	1.303	1.684	2.021	2.423	2.704	3.307

Step 6: Conclusion

With the p-value being greater than 0.200, we know that the p-value is also more than the 0.05 level of significance. Therefore, we fail to reject the null hypothesis. We would conclude, based on our sample data of the 2011 PGA Tour, that there is not enough statistical evidence to conclude that there is a difference in mean scores between the third and fourth round scores for tournament winners on the PGA tour.

In constructing a 95% confidence interval for this difference, we have:

$$\bar{X}_d \pm t * \frac{S_d}{\sqrt{n}} = -0.196 \pm 2.021 * \frac{3.48}{\sqrt{46}} = -0.196 \pm 1.037$$

This results in a 95% confidence interval for the mean difference between third and fourth round scores of PGA Tour winners is from -1.233 to 0.841 strokes. With the interval including zero, the interval would support our conclusion that there is no difference in strokes between the third and fourth rounds for winners on the PGA. Annotated Minitab output is provided in Table 8.7. As it should, the p-value in the output, P-Value = 0.705, falls within our hand-calculated range of the p-value being greater than 0.200.

```
              N    Mean   StDev  SE Mean
Third        46  67.283   2.841    0.419 <-- sample data "group 1"
Fourth       46  67.478   2.364    0.349 <-- sample data "group 2"
Difference   46  -0.196   3.481    0.513 <-- sample data for differences

95% CI for mean difference:(-1.229, 0.838) <-- 95% confidence interval
T-Test of mean difference= 0(vs not = 0): <-- Alternative "not equal"
T-Value = -0.38 P-Value = 0.705
           ^                  ^
     T test statistic      p-value
```

TABLE 8.7: Annotated Minitab output for Example 7—comparing third and fourth round scores for winners on the 2011 PGA Tour.

8.7 Independent versus Dependent Samples

Possibly, by this point in the book, you might be realizing the math involved at this level of statistics is not too terribly difficult. This can be especially true if you are using statistical software or a calculator. Much of the math can be thought of as completing a puzzle, where the formulas represent the puzzle (e.g., test statistic) and the numbers go into their respective positions (e.g., sample mean and standard deviation). However, where difficulty can arise is in *choosing* the correct statistical method for a given situation; i.e., which is the correct puzzle. One such problem area is determining if a research problem involves two independent samples or does the situation reflect paired data. The choice can be straightforward when we recognize, for example, that two measurements are being taken from the same subject, as the case was in Example 7 with golfers on the PGA Tour. However, this identification may not always be so obvious. For one, situations involving family members are frequently treated as paired. If our research was interested, say, in comparing athletic performance of twins or of husbands and sons, the data would be treated as paired. But what about problems where the subjects are not human? Recall in Chapter 2 when we defined a population. We learned that a population does not always have to consist of people.

Consider, for instance, the addition of a second wild card team in major league baseball. Beginning with the 2012 season, MLB will be including a fifth team in the playoffs. Previously, all division winners plus the non-division winner with the best overall record made the playoffs. Now, the second-best non-division winner will also be added to the post season. A question of interest could be if this second wild card team is equal to the first wild card team in terms of ability. That is, would the inclusion of this second team result in adding a team of equal quality, or is this second team of significantly lesser quality? The data gathering is simple enough. The 1995 season was the first year where a wild card team made the play-offs.[4] We would just need to research the number of wins for each wild card team to the team with the next highest number of wins—that is, the team that would have been the second wild card team. The data can be found at www.mlb.com. Beginning from 1996 through 2011, this would represent 17 years and 34 comparisons based on additional teams for both the American and National leagues. We would have as our data two columns with 34 observations in each: one column representing the number of wins for the wild card team, the second column representing the number of wins for the next best team. The question becomes, do we treat this as two independent samples or as paired data?

The answer lies in how you treat the data. Do you feel the two groups are independent? Certainly, you could make the case for this, as no two teams in the same season could fall into both categories. Or does treating the data as paired make more sense, as you can pair the teams by year? The latter design would allow for variability in performance from year to year. For instance, in 2001 the Oakland Athletics were the American League wild card with 102 wins, while the second best non–division winner were the Minnesota Twins with 85 wins. Conversely, in 1999 the Cincinnati Reds won 96 games, which was good enough to be the wild card entry in every year for the National League except one... 1999! That year, the New York Mets won 97 games. Considering this information, the best design would be to consider the *difference* in games won by year: a matched pair design. A better solution would be to treat each league separately to see if there is a difference in the quality of teams in either league.

One final thought on this is how you define a "difference in quality." Number of wins is an obvious choice for quality, but what about difference? Throughout this chapter, we have used zero as the difference, but here that would imply that the teams have the same number of wins. That would not be realistic to assume the second wild card team would have the same number of wins as the first wild card team. In fact, since the inception of the wild card, none of the second wild card teams had the same number of wins as the actual wild card team. In fact, only three times during this period have two teams tied for most wins within a division (2001, 2005, and 2006). With three divisions across the two leagues, there would have been 102 possible chances for this to happen over the 17 years. Having this happen only 3 out of 102, roughly 3% of the time, makes tying seem very unlikely. Therefore, the researcher would need to define what constitutes a reasonable difference.

The choice between independent and paired may not always result in a difference in conclusion, but that is not the point. The idea is to select properly the correct method. To illustrate a situation where the choice of method could result in different results, we return to Example 6, penalties in NHL hockey.

In that example we treated, correctly, the data as paired. What, however, if we mistakenly treated the data as two independent samples? This error can easily occur when using a software package, as the user can unknowingly select the incorrect method. In conducting this comparison and to satisfy a point of the discussion, we will alter our Step 3: Level of Significance from our standard of 0.05 to

4 The 1995 season was shortened due to the 1994 strike.

0.01, another common significance level. Table 8.8 provides the Minitab output for this data, with 8.8a depicting the data as paired and 8.8b as two independent samples. The two p-values, highlighted in red, offer contrasting decisions at a 0.01 level of significance. For paired data, we would not conclude a difference in penalties per game between regular and post season. Conversely, if treated independently, we would conclude a difference exists. If the NHL administrators were using this data to evaluate referee performance, the importance in getting this correct could have terrific ramifications. Decisions made could, for example, affect jobs—some referees could be fired, or call into question the fairness of the game if teams are allowed to play more aggressively during the post season without fear of penalties.

```
8.8a Paired
              N   Mean   StDev   SE Mean
Regular      16  4.431   0.462     0.116
Post         16  3.592   1.069     0.267
Difference   16  0.839   1.228     0.307

99% CI for mean difference: (-0.066, 1.744)
T-Test of mean difference = 0 (vs not = 0):
T-Value = 2.73  P-Value = 0.015

8.8b Two Independent
          N   Mean   StDev   SE Mean
Regular  16  4.431   0.462      0.12
Post     16   3.59    1.07      0.27

Difference = mu (Regular) - mu (Post)
Estimate for difference:  0.839
99% CI for difference:  (0.011, 1.668)
T-Test of difference = 0 (vs not =):
T-Value = 2.88  P-Value = 0.009  DF = 20
```

TABLE 8.8: Minitab output comparing penalties per game during regular and post season in NHL. Table 8.8a treats the data as paired, while Table 8.8b treats the data as two independent samples.

8.8 More on Confidence Intervals, Sample Size, and One-Sided Tests

Confidence Intervals and Level of Significance

As stated in Chapter 7, this book stipulated 0.05 as the level of significance. As a result, the confidence intervals were done at the 95% level. This is the most common significance level, but not the only one. Looking back a section to *Independent versus Paired* and Table 8.8, we used 0.01 as the level of significance. For consistency, this requires us to use 99% confidence intervals. The relationship between alpha, the level of significance, and the confidence level is:

$$\text{Confidence Level} = (1 - \alpha) \times 100\%$$

Therefore, for significance levels of 0.01, 0.05, and 0.10, we would have corresponding confidence levels of 99%, 95%, and 90%. In decision making, comparing the p-value to alpha and interpreting the confidence interval should lead to the same conclusion. You cannot mix and match these two. We return to the NHL penalty paired data example to illustrate this point. Table 8.9 presents the 95% and 99% confidence intervals for the mean difference in per game penalties between regular and post season games in the NHL. Recall the test resulted in a 0.015 p-value.

```
95% CI for mean difference:(0.185, 1.494)

99% CI for mean difference:(-0.066, 1.744)
```

TABLE 8.9: Minitab output of 95% and 99% confidence intervals for mean difference in per game penalties in the NHL.

Upon inspection, we notice that zero is contained in the 99% interval but not the 95% interval. Using the confidence intervals in support of the hypothesis test decision, we would fail to reject the null hypothesis at the 0.01 level of significance (supported by the 99% confidence interval), but reject the null hypothesis at the 0.05 level of significance (supported by the 95% confidence level). If we had mixed these, our decisions would not concur. For instance, with the p-value of 0.015, we would have rejected the null hypothesis at the 0.05 level, but not reached this decision if we used the 99% confidence interval.

Effect of Sample Size in Hypothesis Testing

The size of the sample can have a great effect on decision making. The increased sample size is more likely to lead to a rejection of the null hypothesis, as the larger sample is more likely to detect a small difference as a significant one. That is, the larger the sample, the smaller the p-value. For this reason, when reading the results of a statistical study, one should also consider the confidence intervals and not just the p-value. Practical versus statistical significance should also be considered in large sample studies. On the positive side, increasing sample size also increases the **power** of the test, or the probability of correctly rejecting a false null hypothesis. To demonstrate the effect of sample size on p-value and therefore a statistical decision, consider the coin toss to begin a football game.

A coin is tossed prior to the start of a football game. The team winning the toss gets to choose whether to kick off or receive, or choose which end zone to defend.[5] Understandably, the players, NFL officials, fans, etc., want a fair and balanced coin to be used. However, if not—and a team knew this—they could take advantage of such knowledge. In order to check the fairness the coin toss, say the NFL records how many "heads" come up during the course of the season. They take two random samples of games: one of size 50 and another of size 100. In both samples, heads come up 40% of the time. With a fair toss as the goal, the number of heads should be close to 50%. This makes the hypotheses set up for each sample scenario as:

Ho: p = 0.5 versus Ha: p ≠ 0.5

In the first sample of 50 coin flips with 40% heads (i.e., 20 of the 50 tosses), the result is a z_{stat} of -1.41 and a 0.16 p-value. Using 0.05 as our level of significance, the NFL would conclude that there is no evidence to suggest the coin toss is unfair. They would not reject the null hypothesis.

Yet in the second sample of 100 coin flips with 40% heads (i.e., 40 of the 100 tosses), the result is a z_{stat} of -2.00 and a 0.046 p-value. Using 0.05 as our level of significance, the NFL would conclude that there *is* evidence to suggest the coin toss is unfair. They would reject the null hypothesis.

In both cases, the sample proportion was 40%, but with different decisions based on the size of the sample.

One-Sided Tests from Two-Sided Outcomes

Throughout this chapter we conducted tests with the two-sided, not equal, alternative hypothesis. Early on, however, we mentioned that this may not always be the case. For example, in our discussion on *Independent versus Dependent Samples*, the baseball wild card example would suggest using a one-sided test. This choice makes sense because in comparing number of wins, we know the second wild card team cannot by rule have more wins than the first wild card team. Therefore, using a "not equal" alternative would not be appropriate.

This does not mean we necessarily have to rerun the data. If we have conducted a two-sided test but our interests are in a one-sided outcome, we can just divide the p-value in half. That is, take half of the p-value for a two-sided test to get the p-value for a one-sided test. Just be careful to note two things:

1. If we reject the null hypothesis for the two-sided test, then we would also reject for the one-sided test. This should be clear, since by cutting the p-value in half, we are producing a still smaller p-value. However, if we fail to reject the null hypothesis for the two-sided test, we may or may not reject the null hypothesis for the one-sided test. Depending on the p-value for the two-sided test, taking half may or may not produce a significant result.

2. One must pay attention to how the difference was calculated when going from a two-sided to a one-sided test. In our test of NHL per game penalties, we had a 0.015 p-value for the two-sided

5 The team winning the coin toss can also defer this decision to the second half allowing the loser of the toss to elect how to begin the game.

test. A one-sided p-value would be 0.0075, but we would need to be careful on how this one-sided test was arranged. With our difference coming from *Regular* minus *Post* with a positive difference (see Table 8.6), then the appropriate one-sided test should be:

$$H_o: \mu_d = 0 \text{ versus } H_a: \mu_d > 0$$

Properly set up, we could correctly conclude that on average, more penalties per game are called during the regular season than in the post season.

Expressions and Formulas

1. Test Statistic

Two Proportions

$$Z_{stat} = \frac{(\hat{p}_1 - \hat{p}_2) - 0}{\sqrt{\hat{p}(1 - \hat{p})\left(\frac{1}{n_1} + \frac{1}{n_2}\right)}}$$ where $\hat{p}$ in the denominator is found by:

$$\hat{p} = \frac{\text{Total number of "successes" in both samples}}{\text{Sum of both sample sizes}}$$

This $\hat{p}$ is referred to as a **pooled estimate**, since it "pools together" the total number of successes and sample sizes from the two samples.

Two Independent Means

- If assumption of equal variances IS satisfied:

$$t_{stat} = \frac{(\bar{X}_1 - \bar{X}_2) - 0}{S_p\sqrt{\frac{1}{n_1} + \frac{1}{n_2}}}$$ where is found by:

$$S_p = \sqrt{\frac{(n_1 - 1)S_1^2 + (n_2 - 1)S_2^2}{n_1 + n_2 - 2}}$$

This S_p is referred to as the **pooled estimate**, as it "pools together" the information from the two samples. The degrees of freedom, DF, when using the pooled estimate are found by DF = $n_1 + n_2 - 2$.

- If assumption of equal variances is NOT satisfied:

$$t_{stat} = \frac{(\bar{X}_1 - \bar{X}_2) - 0}{\sqrt{\frac{S_1^2}{n_1} + \frac{S_2^2}{n_2}}}$$

Matched Pairs or Dependent Means

$$t_{stat} = \frac{\bar{X}_d - 0}{\frac{S_d}{\sqrt{n}}}$$

2. Test statistic for two proportions when not pooled, that is Ho is something other than the two proportions are equal.

$$Z_{stat} = \frac{(\hat{p}_1 - \hat{p}_2) - p_0}{\sqrt{\frac{\hat{p}_1(1 - \hat{p}_1)}{n_1} + \frac{\hat{p}_2(1 - \hat{p}_2)}{n_2}}}$$ where p_o is some hypothesized value other than 0.

3. Formula for finding degrees of freedom for unpooled test statistic

$$df = \frac{\left(\frac{S_1^2}{n_1} + \frac{S_2^2}{n_2}\right)^2}{\frac{1}{n_1 - 1}\left(\frac{S_1^2}{n_1}\right)^2 + \frac{1}{n_2 - 1}\left(\frac{S_2^2}{n_2}\right)^2}$$

4. Confidence intervals

Two Proportions

$(\hat{p}_1 - \hat{p}_2) \pm Z^* \sqrt{\hat{p}(1 - \hat{p})\left(\frac{1}{n_1} + \frac{1}{n_2}\right)}$ where $\hat{p}$ is the pooled estimate provided in expression and formula 1.

Two Independent Means

Equal variances assumed: $(\bar{X}_1 - \bar{X}_2) \pm t^* S_p \sqrt{\frac{1}{n_1} + \frac{1}{n_2}}$

Equal variances NOT assumed: $(\bar{X}_1 - \bar{X}_2) \pm t^* \sqrt{\frac{S_1^2}{n_1} + \frac{S_2^2}{n_2}}$

Matched Pairs or Dependent Means

$$X_d \pm t^* \frac{S_d}{\sqrt{n}}$$

Analysis of Variance (ANOVA) 9

At the beginning of Chapter 8, we listed four examples, one of which compared salaries of the NFL, NBA, and MLB. At that time, we mentioned we would save for later the discussion for comparing more than two means. This is later.

From Chapter 8, we learned of the method for testing two means under the assumption of equal variances. This pooled t-test statistic is repeated here:

$$t_{stat} = \frac{(\bar{X}_1 - \bar{X}_2)}{S_p\sqrt{\frac{1}{n_1} + \frac{1}{n_2}}}$$

The t-test statistic is a ratio that compares the differences between the two group means (the numerator) to the variability within the two groups (the pooled standard deviation in the denominator). With the comparison of salaries of these three professional leagues, we now have more than two means to test. One thought might be to run as many two-sample t-tests as the situation permits. For example, we could run tests comparing the NFL to NBA, NFL to MLB, and NBA to NFL. However, such methods do not satisfy our goal of comparing *all* the means together in a single test; instead we are testing each possible pair. By conducting tests in this manner, a chance of making a Type I error applies to each two-mean test. In contrast, conducting one overall test that compares *all* group means, we can control the probability of making a Type I error. This one test, used when we want to compare more than two means, is called Analysis of Variance or ANOVA.

The application of ANOVA methods begins with having a quantitative response variable spread across more than two levels of a categorical explanatory variable. In our professional sports salary example, we have a quantitative response variable—salary, spread across three levels of an explanatory variable, a professional league. The categorical variables in ANOVA are sometimes referred to as **factors** and the different categories within these variables as **factor levels**. When a study has only one factor, as is the case in our example, we refer to this as **one-way ANOVA**. For studies involving two factors, we have **two-way ANOVA**. The focus of this chapter will be on one-way ANOVA.[1]

1 Two-way ANOVA, along with other topics such as non-parametrics and multiple linear regression, can often be special topics in an introductory statistics course. Therefore, they will not be covered in this text.

9.1 One-Way ANOVA

One-way ANOVA, heretofore referred to as simply ANOVA, is used to compare means from at least three groups from one categorical variable. The null hypothesis is that all the population group means are equal versus the alternative that at least one of the population means differs from the others. As with prior chapters, our first step is writing these hypotheses statements.

Step 1: State Hypotheses

Beginning with the null hypothesis we have:

$$Ho: \mu_1 = \mu_2 = \ldots = \mu_g$$

where g is the number of population means being compared. In our above salary example, the null hypothesis would be Ho: $\mu_1 = \mu_2 = \mu_3$ where $g = 3$ for the three leagues. With the test comparing all group means, our alternative hypothesis is simply that not all of the means are equal. This allows for several different ways in which this can be written. Some common ways of writing the alternative hypothesis for ANOVA are:

Ha: at least two of the population means are different.

Ha: at least one population mean differs from the other population means.

Ha: not all of the population means are equal.

You will notice two common threads to the way the alternative is stated. One, whichever you select, the reference is to *population* means; and two, all of the statements only reference a possible difference in at least one mean. Pay special attention to the last one. This alternative does NOT say, "all of the population means differ." Such a statement would be just one of the possible alternatives where at least one mean would differ. Remember, the best we could conclude with an ANOVA test of means is that not all of the means are equal - but this is not the same as saying all of the means are different. That is, if we were to reject Ho, our alternative hypothesis only allows us to conclude a difference in the population means. To better understand this, consider our salary example. Our hypotheses would be written as:

Ho: $\mu_1 = \mu_2 = \mu_3$ versus Ha: at least one population mean differs from the other population means

If we were to reject Ho, this could mean that:

1. All three leagues differ significantly in their population mean salaries, i.e. $\mu_1 \neq \mu_2 \neq \mu_3$, or
2. Any two of the three leagues have the same population mean salary while one is different, i.e.,
   a) $\mu_1 \neq \mu_2 = \mu_3$
   b) $\mu_1 = \mu_2 \neq \mu_3$
   c) $\mu_1 = \mu_3 \neq \mu_2$

Our test in ANOVA will not permit us to make a distinction between how many or which means differ. In Section 9.2, we will discuss methods for determining significant differences between means when we reject our null hypothesis in ANOVA.

Step 2: Check Assumptions

With ANOVA being an extension of the two-mean test with equal variances, the assumptions are quite similar.

- Independent random samples of a quantitative response variable are taken from more than two groups
- The population distribution for the population data of each group is approximately normal
- The population variances of each group are equal[2]

Recall our Rule of Thumb from Chapter 8 for assuming equal variances: If the largest group standard deviation is **not more than double** the smallest group standard deviation, then we deem the variances equal.

Step 3: Set Level of Significance

Remaining consistent with other tests in this text, we set 0.05 as our level of significance.

Step 4: Calculate Test Statistic

Possibly, by this point one might be confused as to why we call this statistical method Analysis of Variance when we are comparing means. The reason is that the test statistic uses evidence about two types of variability. Much like in Chapter 8 where we had a pooled standard deviation, there is also a pooled standard deviation for ANOVA. However, the formula is rather complex, so we will focus on the reasoning behind the formula in order to gain an understanding into the method.

The box plots in Figure 9.1 represent two separate studies, where five salaries (in millions of dollars) are taken from each of the three professional leagues in 2011 (www.cbssports.com): NBA, NFL, and MLB. In analyzing each graph, which case do you think gives stronger evidence that mean salaries for the three leagues are not all the same? From our hypothesis standpoint, we are asking which gives stronger evidence against Ho: $u_1 = u_2 = u_3$? In both studies, the group means are the same. That is, in this example, the mean salary for NBA is 7.8; NFL is 3.8; and MLB is 7.0, represented by the •. What's the difference then? With the means being the same in each study, the variability *between* each pair of sample means is the same. However, the variability *within* each sample is much larger in (a) than in (b),

2 A key component in designing a study that uses ANOVA is to consider having equal sample sizes in each group. The reason for having equal sample sizes is that with such a design, ANOVA test results are robust to violations of normality and/or equal variance assumptions. **Robust** refers to a test still performing sufficiently even in the face of assumption violations.

as shown by the length of the boxes in (a) being larger than in (b). The sample standard deviations in (a) range from 1.6 to 3.7, while in (b) this range is from 0.8 to 1.5. This smaller within-sample variability in (b) is what gives stronger evidence against the null hypothesis in study (b) compared to study (a). This provides two scenarios when we will find stronger evidence against the null hypothesis (that is, when we reject Ho): one, when the within sample variability is small; or two, when the between-sample mean variability is large.

FIGURE 9.1: Box plots for two separate studies, where five salaries from year 2011 are taken from each of the professional leagues: NBA, NFL, and MLB.

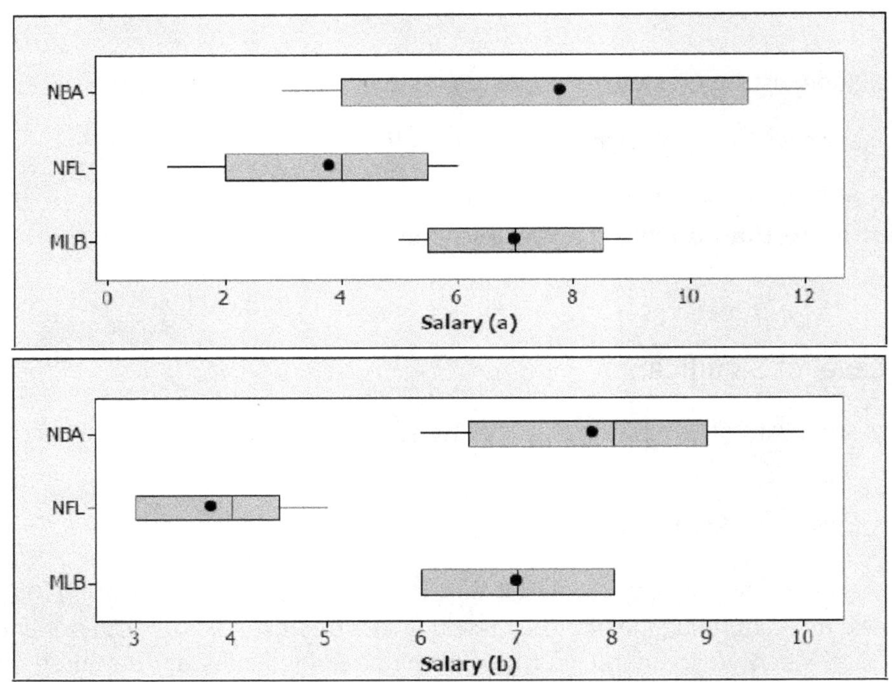

This reasoning, between- and within-group variability, is the focal point of the ANOVA test statistic. We introduce a test statistic from a new distribution called the **F distribution**. This distribution will involve the ratio of two numbers—in the case of ANOVA, two measures of variability. Since these numbers cannot be negative an F test statistic cannot be negative. The shape of the F distribution is also skewed right. The precise shape of the F distribution is determined by two measures of degrees of freedom: one for the numerator and one for the denominator.

As we stated, the ANOVA test is based on comparing the variability between groups to the variability within groups. This sets the **ANOVA test statistic** as:

$$F = \frac{\textit{Variability between groups}}{\textit{Variability within groups}}$$

The degrees of freedom (df) are based on the degrees of freedom used in calculating the variability between groups (numerator df) and those used to calculate the variability within groups (denominator df). If we label these df_1 and df_2, respectively, then these can be found by:

$df_1 = g - 1$ (where g is the number of groups or population means being compared)

$df_2 = N - g$ (where N is the total number sample size, i.e. the sum of the sample sizes in each group)

The calculation of each measure of variability is fairly straightforward when we have equal group sample sizes, such as the cases presented in Figure 9.1, where each group sample size is 5. However, when we have unequal group sample sizes, then each measure of variability is a *weighted* measure, with more weight being given to those sample variances coming from larger groups. Therefore, with *equal* sample sizes, the variability calculations are as follows:

$$\textit{Variability between groups} = \frac{n[(\bar{y}_1 - \bar{y})^2 + (\bar{y}_2 - \bar{y})^2 + ... + (\bar{y}_g - \bar{y})^2]}{g - 1} \quad \text{where,}$$

n = sample size (e.g., with salary example above, this is 5)

$\bar{Y}$ = the overall mean of the sample (e.g., would be the mean of all 15 salaries)

$\bar{y}_g$ = the sample mean for each group (e.g., the sample mean salary for each of the three leagues)

g = number of groups or population means being tested

$$\textit{Variability within groups, } s^2 = \frac{S_1^2 + S_2^2 + ... + S_g^2}{g} \quad \text{where,}$$

S_g^2 = the sample variance of each group (e.g., the square of the sample standard deviations for each of the three leagues)

g = number of groups or population means being tested

Putting these two together results in our F test statistic for ANOVA, which is a ratio of the two population variance estimates,

$$F = \frac{\textit{Estimate of between group population variance}}{\textit{Estimate of within group population variance}}$$

What we will learn is that statistical software will display this ANOVA output. What we see in Table 9.1, is the annotated Minitab output for the data in Figure 9.1(b): The 22.40 is the estimated *between-group variability* and the 1.30 is the estimated *within-group variability*. We can see—as we commented earlier—that a large between-group variability in relation to a small within-group variability produces stronger evidence against the null hypothesis. From the F test statistic above and the output from Table 9.1, we calculate F,

$$F = \frac{\textit{Estimate of between group population variance}}{\textit{Estimate of within group population variance}} = \frac{22.40}{1.30} = 17.23$$

The "SS" column in the output of Table 9.1 relates to the *Sum of Squares*. Starting from the bottom of this column, the **total sum of squares**, which is often noted as SST or SSTo, provides the sum of squares of the whole sample around the overall mean. That is, the SST takes each difference between each observation and the mean of the overall sample, squares these differences, and then sums these squares, i.e., a sum of squares. In the analysis of variance procedure, this total sum of squares is partitioned into the between- and within-sum of squares. This leads to an additive property,

SST = between group SS + within group SS.

From Table 9.1 we can see,

$$60.40 = 44.80 + 15.60$$

What you may notice is that if we divide SST by N - 1, the overall sample size minus 1, we get the sample variance for all the observations if treated as one group. For this example, if we took 60.40 and divided by 15 -1 = 14, the result would be the variance of the 15 sample salaries.

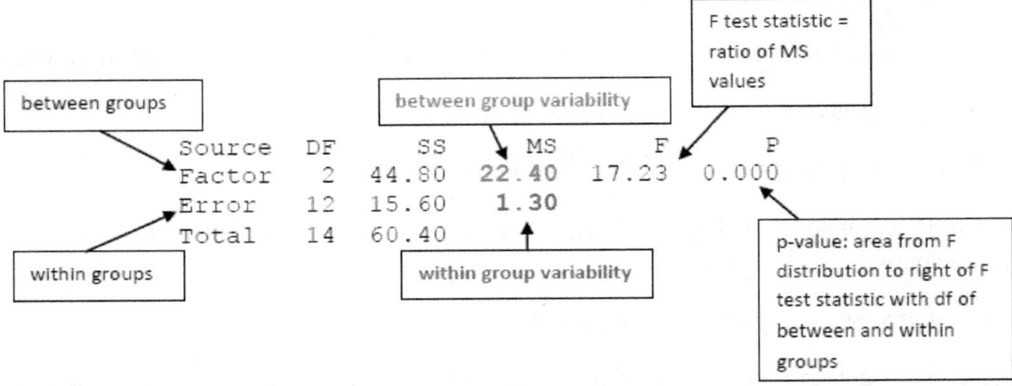

TABLE 9.1: Annotated Minitab output for salary data for Figure 9.1 (b) comparing salaries of five players selected from NBA, NFL, and MLB

Step 5: Calculate p-value

With the *F* test statistic and the two sets of degrees of freedom, df_1 and df_2, we can estimate the range of the p-value from the F-table. The F-table provides a series of F values from the F distribution with a right-tail probability of 0.05, for varying combinations of degrees of freedom. As with reading the T-Table, if the degrees of freedom are not found in the table, a typical practice is to use values closest to—but not exceeding—the correct degrees of freedom.

From our above example, we have an *F* test statistic of 17.13 with df_1 of 2 (found by taking $g - 1 = 3 - 1 = 2$, where g is the number of groups being compared) and df_2 of 12 (found by $N - g = 15 - 3 = 12$, where N is the total size of our sample). For convenience, a portion of the F-Table is provided in Figure 9.2. From the F-table for 2 and 12 degrees of freedom (note that df_1 corresponds to the top row, while df_2 corresponds to the leftmost column), we find an *F* value of 3.89. With our *F* test statistic being *greater* than 3.89 F-value from the table, we conclude that the p-value would be *less* than 0.05. This is supported by the output found in Table 9.1, where we have a three-digit p-value of 0.000.

NOTE: With the F-Table being the third table we addressed, one should have taken notice that as the absolute value of a test statistic increases the area (i.e. probability) to the *right* of that test statistic decreases. As a result, larger test statistics - in absolute value - result in smaller p-values.

FIGURE 9.2: A portion of the F-Table for the ANOVA test of means for the three professional leagues of MLB, NBA, and NFL.

F-Table: F-values for Right Tail Probability = 0.05

	df_1									
df_2	1	2	3	4	5	6	8	12	24	inf.
1	161.45	199.50	215.71	224.58	230.16	233.99	238.88	243.91	249.05	254.31
2	18.51	19.00	19.16	19.25	19.30	19.33	19.37	19.41	19.45	19.50
..	...	...	...	...	...	...	...	...	...	...
..	...	...	...	...	...	...	...	...	...	...
12	4.75	3.89	3.49	3.26	3.11	3.00	2.85	2.69	2.51	2.30

Step 6: Conclusion

As with all our hypothesis tests to this point, our decision is to *reject* the null hypothesis when our p-value is less than our 0.05 level of significance. When we reject Ho in a one-way ANOVA, we are concluding that we have enough statistical evidence to conclude that at least one group *population* mean differs from the remaining group *population* means. Keep in mind, this result would not tell us how many group means or which group means are different. That we will discuss in the next section. Continuing with our Table 9.1 output, with our p-value being less than 0.05, we would reject the null hypothesis and conclude that at least one of the means from the NBA, NFL, and MLB differs from the other means.

If we *fail to reject* the null hypothesis, then we do not have enough statistical evidence to conclude a difference in means. With such a decision, we should avoid the temptation of comparing each possible pair of means. For one, this was not the original research hypothesis—in ANOVA, we are interested in comparing more than two means; and secondly, we introduce the increased probability of making Type I errors, as we mentioned at the start of this chapter.

Example 1:

Since 2002 through 2011, four Major League Baseball teams have arguably been the worst: the Baltimore Orioles, Kansas City Royals, Pittsburgh Pirates, and the Washington Nationals.[3] None of these teams have reached the play-offs nor have they been within less than eight games of making the play-offs. A research question becomes, "Is any one or more of these teams worse than the others?"

Answering this question poses several possibilities, depending on how you want to define *success* (or lack thereof!). One definition could be to use team wins over this 10-year period. Another could be to use games behind in the wild card. An added question could be, Does using either one make a difference? That is, regardless if we use the number of wins or games behind, is the answer the same? To answer these questions, we will consider both wins and games behind, conducting a separate ANOVA for each. However, we will work through them simultaneously, as this will present an opportunity to identify any similarities or differences along the hypothesis testing steps. In Table 9.2, we have the means and standard deviations for wins and games back of the wild card for the four teams (see www.mlb.com). In each case, there is a difference in the means, but the question is are any of these mean differences statistically significant?

Team	Mean Wins	S_{Wins}	Mean Games Back	S_{GB}
Baltimore	69.6	4.03	26.00	5.03
Kansas City	66.8	8.14	28.85	9.49
Pittsburgh	67.9	5.34	23.40	5.75
Washington	72.5	9.18	18.85	8.92

TABLE 9.2: Means and standard deviations of wins and games back of wild card for Baltimore, Kansas City, Pittsburgh, and Washington for the years 2002 through 2011.

Step 1: State Hypotheses

With four teams, we have $g = 4$. Whether we use wins or games behind, the hypotheses are,

Ho: $\mu_1 = \mu_2 = \mu_3 = \mu_4$ versus Ha: at least two of the population means differ.

What could be unclear, though, is which mean represents which team? Does μ_1, for example, represent Pittsburgh? The Nationals? To clarify this, we will use the first letter of the team name to identify the means, rewriting our hypotheses as,

Ho: $\mu_B = \mu_K = \mu_P = \mu_W$ versus Ha: at least two of the population means differ.

Step 2: Check Assumptions

Since our data spans 10 years for each team, we have equal sample sizes: that is, for each of the four teams we have 10 observations. This will help us in the event we have any violations to normality or equal variances. To check for normality, we provide probability plots of each sample. These plots are

3 Prior to moving to Washington, D.C., in 2005, the Washington Nationals were the Montreal Expos.

provided in Figure 9.3 and were computed using Minitab. Recalling from Chapters 7 and 8, when we use probability plots to test for normality, we are interested in p-values being *greater* than 0.05. As Figure 9.3 illustrates, this is true for all samples for wins and games behind. We have satisfied the normal assumption. However, if we had not, by using equal sample sizes we would be helping to protect our test from such a violation. If one does not have access to computer software that can create probability plots, the use of box plots, histograms, etc., can be developed to make judgments regarding the population shapes.

As to the assumption of equal variances, using *Wins* as the response, we notice in Table 9.2 that the largest standard deviation (9.18) is more than double the smallest standard deviation (4.03). However, since we have equal sample sizes, our ANOVA results are robust to this violation. Where *Games Behind* is the response, we can assume equal variances, since the larger standard deviation (9.49) is not more than double the smaller standard deviation (5.03).

FIGURE 9.3: Minitab probability plots for each of the four teams, using wins and games behind wild card. With all p-values exceeding 0.05, we can assume the normal assumption is satisfied.

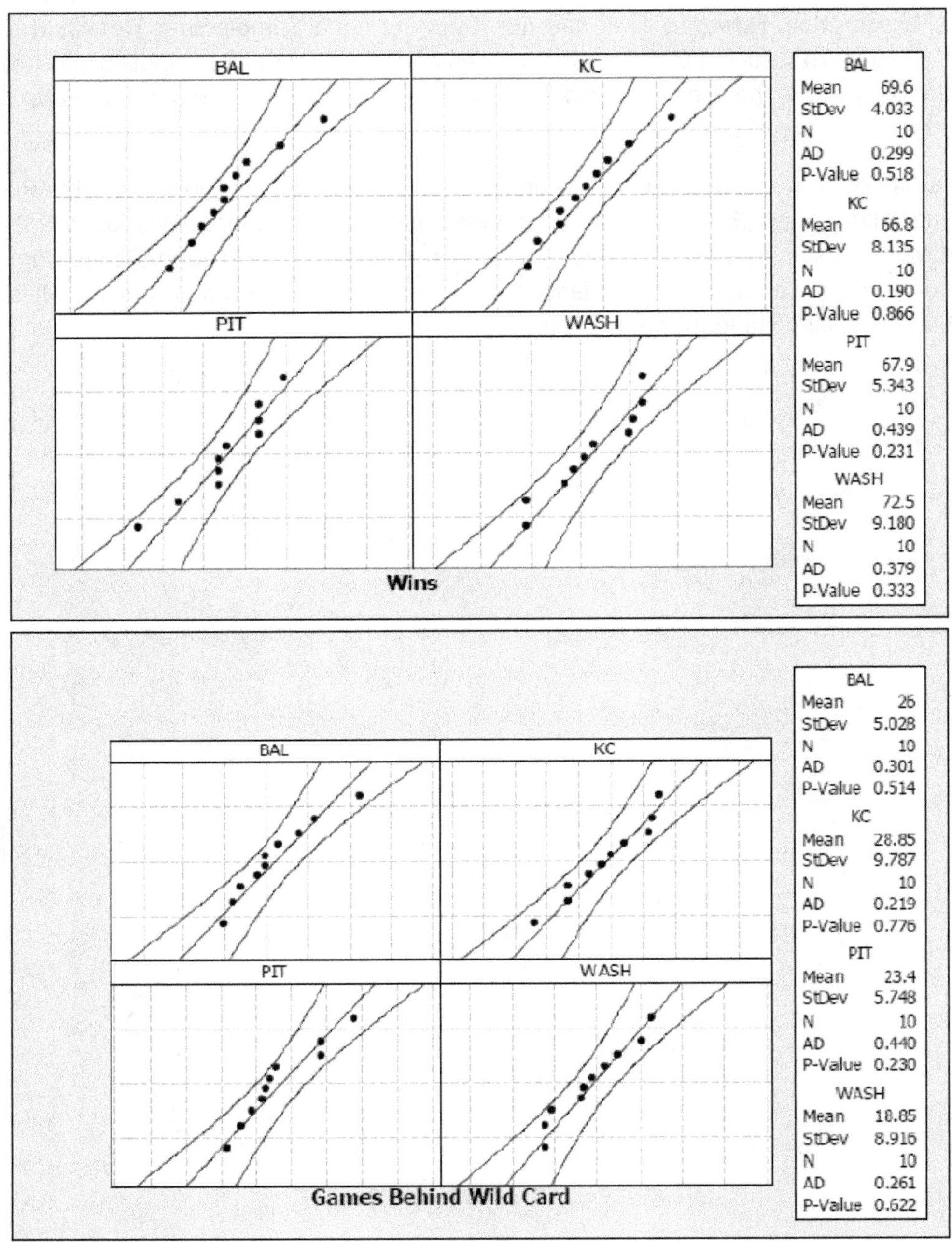

The final assumption is that independent random samples of a quantitative response variable are taken from more than two groups. Most of this is true. The response variable, be it *Wins* or *Games Behind*, are quantitative, and with four teams we have more than two groups. However, what about independent? Yes, these samples were taken over 10 different years, meaning a team cannot have more than one observation from the same year. For example, Pittsburgh only has one win total for 2002, 2003, etc. This is also true for games behind. Yet should we feel comfortable that from one year to the next, a team may remain pretty much the same, barring major losses due to, say, injury? That is—as a typical behavior understanding there might be exceptions—if a team is bad one year, is it safe to say that the team will be bad the following year and vice versa? If one believes this to be true, then independence comes into question, since yearly performance is *dependent* on prior year performance and ANOVA methods may not be a prudent statistical choice. A counterargument in favor of independence is that a bad team, in an effort to improve, should be trying something different from one year to the next. For the sake of argument, we will assume independence, and therefore move forward with our ANOVA test.

Step 3: We Set the Level of Significance at 0.05

Step 4: Calculate Test Statistic

Given the complexity in calculating the *F* test statistic, we will analyze computer software to assist us. In Table 9.3, we have Minitab output for both studies.

Source	DF	SS	MS	F	P
Factor	3	185.0	61.7	1.26	0.302
Error	36	1757.4	48.8		
Total	39	1942.4			

Response = Wins (Study A)

Source	DF	SS	MS	F	P
Factor	3	541.0	180.3	3.09	0.039
Error	36	2102.5	58.4		
Total	39	2643.5			

Response = Games Behind (Study B)

TABLE 9.3: Minitab ANOVA output for Wins (Study A) and Games Behind (Study B) of the four Major League Baseball teams: Baltimore, Kansas City, Pittsburgh, and Washington.

Upon closer inspection of Table 9.3, the *F* test statistics are calculated by,

$$(Wins): F = \frac{Estimate\ of\ between\ group\ population\ variance}{Estimate\ of\ within\ group\ population\ variance} = \frac{61.7}{48.6} = 1.26$$

$$(Games\ Behind): F = \frac{Estimate\ of\ between\ group\ population\ variance}{Estimate\ of\ within\ group\ population\ variance} = \frac{180.3}{58.4} = 3.09$$

What this example provides is an excellent illustration of how the changes in the between- and within-group variability affects our test statistic. In either study, we have the same degrees of freedom: df_1 =

$g - 1 = 4 - 1 = 3$ and $df_2 = N - g = 40 - 4 = 36$. Yet the ratio of between and within group variability is much smaller for *Wins* than for *Games Behind*. This produces a smaller *F* test statistic for *Wins*, resulting in a larger p-value, in turn making rejecting the null hypothesis more difficult.

Step 5: Calculate p-value

From Step 4, we listed the test statistic and degrees of freedom for both response options. These were,

Wins: $F = 1.26$ with $df_1 = 3$ and $df_2 = 36$ and *Games Behind*: $F = 3.09$ with $df_1 = 3$ and $df_2 = 36$

From the F-Table, a portion of which is provided in Figure 9.4, we use $df_1 = 3$ and df_2 of 30 since 36 is not listed, we have the following range of p-values for these test statistics:

Wins: Since $F = 1.26$ is *less* than the F-Table value of 2.92 for df_1 of 3 and df_2 of 30, we conclude that the p-value for this test would be *greater* than 0.05. This is confirmed in the Minitab output of Table 9.3 with the p-value of 0.302 for Study (A).

Games Behind: Since $F = 3.09$ is *greater* than the F-Table *F* value of 2.92 for df_1 of 3 and df_2 of 30, we conclude that the p-value for this test would be *less* than 0.05. This is confirmed in the Minitab output of Table 9.3 with the p-value of 0.039 for Study (B).

FIGURE 9.4: A portion of F-Table for Example 1.

F-Table: F-values for Right Tail Probability = 0.05

	df_1									
df_2	1	2	3	4	5	6	8	12	24	inf.
1	161.45	199.50	215.71	224.58	230.16	233.99	238.88	243.91	249.05	254.31
2	18.51	19.00	19.16	19.25	19.30	19.33	19.37	19.41	19.45	19.50
..	...	...	...	...	...	...	...	...	...	...
..	...	...	...	...	...	...	...	...	...	...
30	4.17	3.32	2.92	2.69	2.53	2.42	2.27	2.09	1.89	1.62

Step 6: Conclusion

This is where our ANOVA tests get interesting! On one hand, with *Wins* we would fail to reject the null hypothesis, since our p-value is greater than 0.05. We cannot say any of the mean number of wins for these four teams is significantly different from another mean. As a result, we do not have statistical evidence that shows any team or teams is/are worse than another team in terms of average wins per season.

On the other hand, with *Games Behind* we would reject the null hypothesis, as our p-value is less than 0.05. With this response variable, we can conclude that a statistical difference exists between at least two of the means. That is, at least one team can be considered worse than the other teams (or alternatively, one team could be considered better than the other three teams). The use of two different definitions of "success" presents two conflicting conclusions: Which one is correct?

Actually, both results are correct for the data used. The real question, as we stated at the beginning of this example, is in *how* you want to define success. With two different definitions, we ended up with two separate and different results. With the non-rejection using *Wins*, we are finished with the study; there is no continuation, since we decided no statistical differences exist. However, with the statistically significant result (i.e., rejecting the null hypothesis) coming from using *Games Behind*, a follow-up investigation should take place to determine the cause of this rejection. In other words, we need to determine which mean, or means, are different. As of now, by rejecting the null hypothesis, we can only conclude there is a difference in some of the means. We just do not know how many means differ, which means differ, nor how large is this difference. To address this issue when we have an ANOVA *F*-test rejection, we examine confidence intervals that compare all possible pairs of means.

9.2 Following Up a Rejection of an ANOVA F-Test

With a rejection in an ANOVA test, our follow-up is which mean or means differ. The investigation takes place by calculating confidence intervals for all possible combination of pairs of means. In our example using the four MLB teams, we have *six* possible pairs of means to compare. These are:

Baltimore with Kansas City

Baltimore with Pittsburgh

Baltimore with Washington

Kansas City with Pittsburgh

Kansas City with Washington

Pittsburgh with Washington.

In general, for g groups there are $g(g-1)/2$ pairs. With our example using $g=4$, this comes to $4(4-1)/2 = 6$ pairs.

As we learned in Section 9.1, the denominator in our *F*-test statistic is an estimate of the within-group variance. The value of this can be found in the ANOVA output under the column MS in the row marked Error. By taking the square root of this value. we have an estimate of the pooled standard deviation. We can use this estimate to serve as the pooled standard deviation in calculating differences between two independent means as we did in Chapter 8. The difference being that we now have multiple pairs of means to compare, thus multiple confidence intervals to construct. The general formula for constructing 95% confidence intervals for these pairwise differences in means is,

$$(\bar{y}_i - \bar{y}_j) \pm t^* s\sqrt{\frac{1}{n_i} + \frac{1}{n_j}} \text{ where } i \neq j.$$

The t-value from the T-Table, a portion of which is provided in Figure 9.5, using the column for 95% confidence interval (right-tail probability of 0.025) with degrees of freedom (df) equal to N - *g*, with N representing the total sample size and *g* the number of groups or means being compared. The *i* and *j* subscripts represent any two of the groups, just not the same two groups (this explains the $i \neq j$). Recalling in Chapter 8 what we learned about confidence intervals where we compare two groups, a difference in the groups being compared is represented by an interval that does not contain 0. As an example of applying this confidence interval formula, we will select only two of the possible six intervals: one where we compare the two closest sample means, and another comparing the two most distant sample means from the data in Table 9.2.

FIGURE 9.5: Partial T-Table indicating multiplier for 95% confidence intervals for comparison of mean salary data for MLB, NBA, and NFL.

T-Table: t Distribution Confidence Interval and Critical Values

	Confidence Level					
	80%	90%	95%	98%	99%	99.8%
	Right Tail Probability					
df	$t_{0.10}$	$t_{0.05}$	$t_{0.025}$	$t_{0.01}$	$t_{0.005}$	$t_{0.001}$
1	3.078	6.314	12.706	31.821	63.657	318.289
2	1.886	2.920	4.303	6.965	9.925	22.328
..	...	...	...	...	...	...
..	...	...	...	...	...	...
35	1.306	1.690	2.030	2.438	2.724	3.340

The two closest sample means are Baltimore and Kansas City with respective means *Games Behind* of 26.00 and 28.85. From the output in Table 9.3, the pooled standard deviation can be found by taking the square root of 58.4, resulting in $s = 7.642$. With df = 40 - 4 = 36, we use the df row of 35 in our T-Table to get a 95% confidence interval t-value of 2.030. With each group sample size being 10, we construct our interval as follows,

$$(\bar{y}_B - \bar{y}_K) \pm t^* s\sqrt{\frac{1}{n_B} + \frac{1}{n_K}} = (26.00 - 28.85) \pm 2.030 * 7.642\sqrt{\frac{1}{10} + \frac{1}{10}} = -2.85 \pm 6.94$$

The result is a 95% interval for the mean difference in *Games Behind* between Baltimore and Kansas City ranging from -9.79 to 4.09. With this interval including 0, we cannot conclude a difference in means between Baltimore and Kansas City.

The two sample means with the biggest difference are Kansas City and Washington with respective means *Games Behind* of 28.85 and 18.85. From the output in Table 9.3, the pooled standard deviation remains as $s = 7.642$. We still have df = 40 - 4 = 36, using the df row of 35 in our T-Table to get a 95% confidence interval t-value of 2.030. With each group sample size being 10, the only difference is the two sample means. We construct our interval as follows,

$$(\bar{y}_B - \bar{y}_K) \pm t^* s\sqrt{\frac{1}{n_B} + \frac{1}{n_K}} = (28.85 - 18.85) \pm 2.030 * 7.642\sqrt{\frac{1}{10} + \frac{1}{10}} = 10.00 \pm 6.94$$

The result is a 95% interval for the mean difference in *Games Behind* between Kansas City and Washington ranging from 3.06 to 16.94. With this interval *not* including 0, we can conclude a difference in means between Kansas City and Washington. Paying special attention to *how* this difference in means was calculated, we see can say, on average, that Kansas City ends the season significantly more games behind the wild card team than Washington. Furthermore, Kansas City has performed statistically worse than Washington when using the metric *Games Behind* as a measure of performance. We now know that at least these two means are different statistically, but we still do not know if any or all of the remaining mean pairs differ. We would have to complete the remaining confidence intervals.

There is one drawback to this method of calculating each pairwise mean 95% confidence interval: we are only 95% sure that any *specific* confidence interval we built correctly contains the true estimated difference. For instance, we are 95% confident that the Baltimore-Kansas City interval is correct, as we are 95% confident in the Kansas City-Washington confidence interval. However, we are *not* 95% confident that *both* intervals correctly contain the true difference. In order to effectively evaluate all possible paired mean intervals—that is, to conclude we are 95% confident that *all* of the intervals correctly capture the true difference—we use methods called **multiple comparisons**. These methods adjust the formula to allow one to draw conclusions that apply simultaneously to the entire set of confidence intervals. These are sometimes referred to as **simultaneous confidence intervals**.

Several multiple comparison methods exist, each having advantages and disadvantages. For the purpose of our discussion, we will use the **Tukey method** of multiple comparisons. The advantage is that the Tukey method produces an overall confidence level that is quite close to the desired level (e.g., 95%), but the procedure is complicated to construct by hand. Therefore, we will use software to perform these calculations. These results are displayed in Table 9.4.

Difference	Individual CI	Tukey Multiple Comparisons CI
Baltimore-Kansas City	(-9.78 to 4.08)	(-12.06 to 6.36)
Baltimore-Pittsburgh	(-4.33 to 9.53)	(-6.61 to 11.81)
Baltimore-Washington	(0.22 to 14.08)*	(-2.06 to 16.36)
Kansas City-Pittsburgh	(-1.48 to 12.38)	(-3.76 to 14.66)
Kansas City-Washington	(3.07 to 16.93)*	(0.79 to 19.21)*
Pittsburgh-Washington	(-2.38 to 11.48)	(-4.66 to 13.76)

TABLE 9.4: 95% confidence intervals for each pairwise difference in means for the four poorest MLB teams from 2002 through 2011. The table presents both individual and Tukey multiple comparison intervals. The * indicates significant differences between that specific mean pair by 0 not being captured by the interval.

What we present in Table 9.4 are 95% confidence intervals using the individual confidence interval formula and simultaneous confidence intervals using Tukey's method of multiple comparisons. Intervals marked with an * indicate an interval where a significant difference exists between the means based on the interval not containing 0. As we can see, we have two such intervals using the individual method—statistical difference between Baltimore-Washington and Kansas City-Washington, but only one such interval using Tukey's method—between Kansas City-Washington. Notice as well that the Tukey intervals are *wider* than the individual intervals. This occurs as each separate interval calculated using the Tukey method uses a higher level of confidence in order for the result to be at our desired 95% level for all intervals simultaneously. Intuitively, this should make sense, as one should expect being correct on a set of intervals to be more difficult in being correct on any one interval. Thinking of it another way, if you were to bet on any one interval being correct or all six being correct, which of these bets would you have more confidence in winning? Hopefully, you selected being correct on any one interval, as this would be easier to select compared to being right on all six.

In conclusion, the research does not indicate any one team from these four as the worst team. However, one can make an argument that based on the results of these 10 years, the Washington Nationals were the best of these four teams.

Which interval method to use, individual or simultaneous, mostly comes down to how many groups—and therefore possible pairwise mean confidence intervals—there are to construct. The more groups, *g*, you have in your analysis, the more pairs of means to compare. If *g* is three, there are 3(3-1)/2 = 3 pairwise mean differences. With *g* of 6, we would have 6(6-1)/2 = 15 pairwise mean differences. As this number grows, so too does our probability that at least one of these confidence intervals will be wrong in estimating the true difference in means. More to the point, the more intervals constructed, the greater our chances that at least one interval is in error in the estimation. We can control for this error rate by using multiple comparison methods. Unfortunately, there is not a defined *g* number of groups at which one should automatically know to use multiple comparison methods.

Expressions and Formulas

1. One-way ANOVA hypotheses

   Ho: Ho: $\mu_1 = \mu_2 = \ldots = \mu_g$ where *g* is the number of group means being compared

   Ha: at least two population means are different

2. *F*-test statistic of one-way ANOVA

$$F = \frac{\text{Variability between groups}}{\text{Variability within groups}} = \frac{\text{MS factor}}{\text{MS error}}$$

$df_1 = g - 1$ (where *g* is the number of groups or population means being compared)

$df_2 = N - g$ (where N is the total number sample size, the sum of the sample sizes in each group

3. Variance estimates for *F*-test statistic when group sample sizes are equal

$$\text{Variability between groups} = \frac{n[(\bar{y}_1 - \bar{y})^2 + (\bar{y}_2 - \bar{y})^2 + \ldots + (\bar{y}_g - \bar{y})^2]}{g - 1}$$ where,

n = sample size (e.g., with salary example above. this is 5)
$\bar{y}_g$ = the overall mean of the sample (e.g.. would be the mean of all 15 salaries)
$\bar{Y}$ = the sample mean for each group (e.g., the sample mean salary for each of the three leagues)

g = number of group or population means being tested

Variability within groups, $s^2 = \frac{S_1^2 + S_2^2 + + S_g^2}{g}$ where,

= the sample variance of each group (e.g., the square of the sample standard deviations for each of the 3 leagues)

g = number of group or population means being tested

4. Individual confidence interval for making any two-mean comparison when rejecting Ho

$(\bar{y}_i - \bar{y}_j) \pm t * s\sqrt{\frac{1}{n_i} + \frac{1}{n_j}}$ where $i \neq j$ and S equals the square root of MSE.

Comparing Two Categorical Variables 10

Before moving on, we will take a moment to briefly review the statistical inference methods we have covered up to this point. We began with confidence intervals for one proportion and one mean in Chapter 6. In Chapter 7, we introduced ourselves to hypothesis testing, starting with one mean and one proportion. We expanded on these concepts in Chapter 8 when we examined two proportions and two means—paired and independent. This also included confidence intervals for these comparisons. Lastly, in Chapter 9, we extended our comparisons of means to include situations where we had more than two means when we learned of analysis of variance techniques. Now in Chapter 10, we will investigate methods for determining if there is statistical evidence of an association between two categorical variables. Until we have such evidence, our assumption is that the two categorical variables are not associated; that is, they are independent. Another way to think of association is dependence: Is one variable dependent on some other variable? Way back in Section 2.2, we mentioned association in terms of a response variable being associated with some explanatory variable. We will refine this to our current discussion, whereby both the response and explanatory variables are categorical. For example, medical scientists are currently interested in studying whether retired NFL players suffer a higher rate of depression compared to the general population. Such a study would have two categorical variables:

1. NFL experience: Did the subject play in the NFL (yes or no responses); and
2. Depression: Does the subject suffer from depression (again, yes or no responses).

In these studies, the response variable is *depression* with *NFL experience* acting as the explanatory variable. The best place to begin is with an example that illustrates how data may appear when independent and dependent.

10.1: Independence versus Dependence

The most common and best method for displaying data from two categorical variables is to use what is called a **contingency table**. A contingency table displays as rows the levels of one categorical variable and as columns the levels of a second categorical variable. A typical practice is to us the explanatory variable for the rows and the response variable for the columns. The table is then populated with the number of sample observations for each combination of these rows and columns called the **observed counts**. Each location of a count in a contingency table is referred to as a **cell.** For example, if two categorical variables each had two levels, then the resulting table would consist of four cells: one for each combination of the two levels for the categorical variables.

Consider LeBron James's shooting ability from two- and three-point range. During the 2011–2012 NBA regular season, James made 53.1% of his shots from the floor. From his season totals (see www.nba.com), can we say if there was an association between type of shot—two- or three-pointer—and his shooting performance? In Table 10.1,[1] we have a contingency table that displays James's shot totals for the 2011–2012 season.

	Shot Result		
Shot	Made	Miss	**Total**
Two pt	567	453	**1020**
Three pt	54	95	**149**
Total	**621**	**548**	**1169**

TABLE 10.1: Contingency table of LeBron James's shot results during the 2011–2012 NBA regular season.

Although using observed counts are helpful, they are not that visually informative in determining if an association exists between the two variables. With different totals for the outcome categories, the counts can be misleading as to whether an association exists. Both cell counts for "Two pt" exceed the cell counts for "Three pt," which can potentially give the impression that an association exists.

A better format for displaying data for two categorical variables is to use **conditional percentages** in the table instead of observed counts. These conditional percentages refer to the percentage that each row cell count is of that row's total. By using this format, we can compare the percentages across the rows of the explanatory variable. If these percentages are similar, then this would indicate no association; if dissimilar, then a possible association exists. In our example, the conditional percentages for the outcome "Two pt" are calculated by the row counts for "Two pt" being divided by 1020, the total number of two-point shots taken by James. Similarly, the conditional percentages for the outcome "Three pt" is calculated by row counts for "Three pt" being divided by 149, the total number of three-point shots taken by James. Each of these is then be multiplied by 100% to get percentages. Table 10.2 displays these conditional percentages.

	Shot Percentage	
Shot	Made	Miss
Two pt	55.6%	44.5%
Three pt	36.2%	63.8%

TABLE 10.2: Conditional percentages of LeBron James's shot results during the 2011–2012 NBA regular season.

1 When a table displaying two categorical variables, where both variables have two levels as in Table 10.1, we refer to the *size* of this table as a **two-by-two table**. In general, the size of a contingency table is the number of rows, **r**, by the number of columns, **c** , and these are referred to as **two-way tables**.

As one can see with the conditional percentages in Table 10.2, James had a higher made percentage for two-pointers compared to three-pointers. This would give an indication that there is an association (dependence) between the two categorical variables of shot type and shot percentage. If these conditional percentages were identical, then the two variables would be independent. However, despite these percentages not being identical, we cannot conclude *statistically* if there is an association. To do this, we use a **Chi-Square Test of Independence** to test for an association between two categorical variables.

10.2: The Expected Table of Counts

How can we make a judgment regarding whether or not two categorical variables are independent or dependent? As we saw in Table 10.2, LeBron James had different shooting percentages depending on what type of shot he took, but were these percentages different simply due to sampling error, or were they too unusual to be the result of chance?

What we have in any contingency table are the **observed counts**; another term for contingency table is **observed table**. These are the counts, or data, provided to us by our sample. For instance, to get the data in Table 10.1 we would go to our resource, www.nba.com, to observe how many shots LeBron took and made as classified by two- and three-point shots. The chi-square test is designed to use these observed counts to determine if the differences are due to sampling error—thus, are the two variables independent, or is what we observed evidence that the two variables are dependent? However, the observed counts alone cannot tell us if these differences are due to chance or not. We need to have something to compare these counts to. What we need is a table that would provide us the counts if the two variables were *independent*. In other words, we want a table that would distribute the observed data in such a way as to represent how the counts would appear if the variables were independent. This would be a table of **expected counts**, also referred to as an **expected table**.

The calculation of expected counts is as follows:[2]

FIGURE 10.1: Process for finding expected counts for a two-by-two table.

	Column 1	Column 2	Total
Row 1	$\frac{(R1 * C1)}{OT}$	$\frac{(R1 * C2)}{OT}$	Row 1 Total (R1)
Row 2	$\frac{(R2 * C1)}{OT}$	$\frac{(R2 * C2)}{OT}$	Row 2 Total (R2)
Total	Column 1 Total (C1)	Column 2Total (C2)	Overall Total (OT)

2 The calculations shown are for a two-by-two table, but the logic would extend for any size table. The expected count for each table cell would be found by taking that cell's row total times column total and dividing by the overall total.

The table of expected counts represents how the observed data would appear if the two variables were independent. This table is developed using each row and column total in comparison with the overall total. If we apply the calculations in Figure 10.1 to the contingency table of observed counts in Table 10.1, we produce the expected table found in Table 10.3.

	Column 1	Column 2	Total
Row 1	$\frac{(1020 * 621)}{1169} = 541.8$	$\frac{(1020 * 548)}{1169} = 478.2$	**1020**
Row 2	$\frac{(149 * 621)}{1169} = 79.2$	$\frac{(149 * 548)}{1169} = 69.8$	**149**
Total	**621**	**548**	**1169**

TABLE 10.3: Expected cell counts for LeBron James's shot results during the 2011–2012 NBA regular season.

With these two tables, one with observed counts (see Table 10.1) and the other with expected counts (see Table 10.3), we compare the cell counts. Assuming the two variables are not related (independent), we would expect to find the cell counts in the observed table to be close in number to those cell counts in the expected table. Note that the expected cell counts will sum to the same row and column totals for that found in the observed table (allowing for possible rounding error). The chi-square test will compare statistically these two tables. The results will provide statistical evidence to conclude if the differences in counts are small enough to be due to random error (independent), or if the differences in counts are large enough to indicate an association between the variables (dependence). If the conclusion is in favor of dependence, we would say that there is a relationship, or an association, between the two categorical variables.

10.3: Chi-Square Test of Independence between Two Categorical Variables

Step 1: State Hypotheses

Similar to our ANOVA *F* test, our hypothesis for our chi-square test can be written in text form. As the test assumes the variables are not related until there is statistical evidence to say otherwise, the null and alternative hypothesis are written,

Ho: In the population, the two variables are independent.

Ha: In the population, the two variables are dependent (associated).

Another common way to state these hypotheses is,

Ho: In the population, there is no relationship between the two variables.

Ha: In the population, there is a relationship between the two variables.

Step 2: Check Assumptions

- The data represent random samples of two categorical variables
- At least 80% of the expected cell counts are at least five[3]

Step 3: Set Level of Significance (per prior tests in this book, we have fixed this at 0.05)

Step 4: Calculate Test Statistic

The test statistic is developed based on our discussion in Section 10.2, where we stated the chi-square test will compare the observed and expected tables. The chi-square test statistic is,

$$X^2 = \sum \frac{(Observed\ Count - Expected\ Count)^2}{Expected\ Count} = \sum \frac{(O - E)^2}{E}$$

The symbol *X* is the capital Greek letter chi, and the statistic compares the count in each cell of the observed table to its cell counterpart in the expected table. For shorthand, we can use "O" to represent *Observed* Count and "E" to represent *Expected Count*. Since these differences in cell counts are squared, the **chi-square test statistic cannot be negative**. Similar to the *F* test statistic for ANOVA in Chapter 9, the minimum chi-square value is 0. And when would this occur? When the observed and expected counts are identical; we have complete independence.

Step 5: Find p-value

The chi-square distribution takes on a skewed distribution, where the precise shape is based on the degrees of freedom. The degrees of freedom for a table displaying two categorical variables are based on the number of rows, *r*, and columns, *c*, of that table—excluding those representing the total. The **degrees of freedom** for a table are found by,

$$df = (r - 1) \times (c - 1)$$

From our LeBron James example, since we have two rows ($r = 2$) and two columns ($c = 2$), the degrees of freedom would be $(2 - 1) \times (2 - 1) = 1$. The Chi-Square Table in the appendix gives chi-square values for various combinations of degrees and freedom and right-tail probabilities. One might notice that the Chi-Square Table and the T-Table are similarly presented, with the exception that we can have negative t-values. This makes reading the Chi-Square Table to get a p-value similar to reading the T-Table to get a p-value.

3 When applying this assumption to a two-by-two table which always consists of four cells, it implies that all expected cell counts be at least five.

Step 6: Conclusion

As with all of our tests to this point, our decision is to *reject* the null hypothesis when our p-value is less than our 0.05 level of significance. When we reject Ho in a chi-square test of independence, we conclude that there is enough evidence to claim that the two variables are related in the *population*. When we make a rejection using a chi-square test of independence, we would want to follow up by inspecting each cell's contribution to the calculation of the chi-square test statistic. Cell or cells making large contributions would indicate the levels of the two categorical variables offering the greatest departure from independence.

Example 1: LeBron James (continued)

We continue with the LeBron James data to finish what we started. We put what we have done up to this point within the confines of our hypothesis steps.

Step 1: State Hypotheses

Ho: In the population, shot type and shot result are independent for LeBron.

Ha: In the population, shot type and shot result are dependent (associated) for LeBron.

Note: From a non-statistical point of view, we are saying with our null hypothesis that regardless if he takes a two- or three-point shot, LeBron has the same rate of success.

Step 2: Check Assumptions

We will assume the data for 2011–2012 represents a random sample of the population of all shot results for LeBron. The two variables, *shot type* and *shot result*, are categorical. From the expected table (Table 10.3), we see that expected counts in each cell are all at least five.

	Column 1	Column 2	Total
Row 1	$\frac{(1020 * 621)}{1169} = 541.8$	$\frac{(1020 * 548)}{1169} = 478.2$	**1020**
Row 2	$\frac{(149 * 621)}{1169} = 79.2$	$\frac{(149 * 548)}{1169} = 69.8$	**149**
Total	**621**	**548**	**1169**

TABLE 10.3 (Repeated): Expected cell counts for LeBron James's shot results during the 2011–2012 NBA regular season.

Step 3: We Set the Level of Significance at 0.05

Step 4: Calculate Test Statistic

Table 10.4 displays both the observed and expected counts for the LeBron James shot data.

	Observed		Expected	
	Shot Result		Shot Result	
Shot	Made	Miss	Made	Miss
Two pt	567	453	541.8	478.2
Three pt	54	95	79.2	69.8

TABLE 10.4: Observed and expected counts for the LeBron James shot data in Example 1.

Substituting the data from Table 10.4 into our chi-square test statistic formula (when calculating by hand, remember to *square* each difference in the numerator),

$$X^2 = \sum \frac{(Observed\ Count - Expected\ Count)^2}{Expected\ Count} = \frac{(567-541.8)^2}{541.8} + \frac{(453-478.2)^2}{478.2} + \frac{(54-79.2)^2}{79.2} + \frac{(95-69.8)^2}{69.8}$$

$$= 1.17 + 1.32 + 7.99 + 9.06 = 19.54$$

Degrees of freedom (df) = $(r-1) \times (c-1) = (2-1) \times (2-1) = 1$

Step 5: Calculate p-value

With a chi-square test statistic of 19.54 and degrees of freedom of 1, we turn to the chi-square table to find the p-value. Figure 10.2 provides a portion of the Chi-Square Table, with the row for degrees of freedom of one highlighted.

FIGURE 10.2: First four degrees of freedom chi-square values for various right-tail probabilities.

Chi-Square Table: Chi-Square (X^2) Values for Various Right Tail Probabilities

	Right Tail Probability						
df	0.250	0.100	0.050	0.025	0.010	0.005	0.001
1	1.32	2.71	3.84	5.02	6.63	7.88	10.83
2	2.77	4.61	5.99	7.38	9.21	10.60	13.82
3	4.11	6.25	7.81	9.35	11.34	12.84	16.27
4	5.39	7.78	9.49	11.14	13.28	14.86	18.47

We notice in the table that as we go from left to right, the chi-square values *increase* as the right-tail probabilities *decrease*. Going across that row of df = 1, we find that our chi-square value of 19.54 *exceeds* the last chi-square value of 10.83. This means that our p-value is *less than* 0.001.

Step 6: Conclusion

With our p-value being less than 0.001, the p-value would also be less than 0.05. We reject the null hypothesis and conclude that there is enough statistical evidence to support stating there is a relationship in the population for *shot type* and *shot result* for LeBron James. Table 10.5 provides highlighted Minitab output for our test. Since we rejected the null hypothesis, we consider the chi-square contributions made by each cell to ascertain which levels of the variables might be supporting this relationship. Recall that a contribution value of zero indicates complete independence, as this would only occur when the observed and expected cell counts are equal. The larger a chi-square contribution is, the more likely the dependence rests with those variable levels. From Table 10.5, we can see that the chi-square contributions are quite close to zero for two-point shots. This is not surprising, given that the observed and expected cell counts in Table 10.4 were fairly close for LeBron's two-point shots. The large chi-square contributions for three-point shots, however, give visual evidence that the reason for the relationship between shot type and shot result is due to LeBron performing below what was expected under the null hypothesis when shooting three-pointers. Our conclusion is probably not that surprising, however, as we probably would have expected LeBron to perform better when shooting two-pointers as opposed to shooting three-pointers.

```
Observed counts listed first
Expected counts are printed below observed counts
Chi-Square contributions are printed below expected counts

              Made     Miss   Total
Two Pt         567      453    1020
            541.85   478.15
             1.168    1.323

Three Pt        54       95     149
             79.15    69.85
             7.993    9.057

Total          621      548    1169

Chi-Sq = 19.541, DF = 1, P-Value = 0.000
```

TABLE 10.5: Highlighted Minitab output for chi-square test of LeBron James data in Example 1.

Example 2: Is There a Relationship between "Pitch Count" and "Bat Outcome" for Alex Rodriguez?

We wish to examine if Alex Rodriguez, "A-Rod," performs differently when facing different pitch count situations. Three levels to *pitch count* are evaluated: *ahead* (more balls than strikes), *behind* (more strikes than balls), and *even* (equal numbers of balls and strikes). The variable has two *outcomes*:

Rodriguez either gets a *hit* or makes an *out*.[4] Since in the analysis *pitch count* will explain *outcome*, the rows will be *pitch count*, and columns the *outcome*. This will result in a table with 3 rows and 2 columns; a three-by-two table.

Step 1: State Hypotheses

Ho: In the population, pitch count and bat outcome are independent for A-Rod.

Ha: In the population, pitch count and bat outcome are related (dependent) for A-Rod.

Step 2: Check Assumptions

We will assume the data from the 2011 season represents a random sample of the population of all at-bat outcomes for pitch count for A-Rod. The two variables, *pitch count* and *bat outcome*, are categorical. Since we have six cells, recall our table is now a 3 x 2, the assumption that at least 80% of the expected cell counts are at least five implies that at least five of the six cells in the expected table have counts of five or more. From www.mlb.com, we get the 2011 observed count data for A-Rod, which we use to calculate the expected counts. This data is displayed in Table 10.6. Since all six cells in the expected table have counts of at least five, we have satisfied the assumption.

OBSERVED:

	Bat Outcome		
Pitch Count	Hit	Out	**Total**
Ahead	44	96	**140**
Behind	22	82	**104**
Even	37	92	**129**
Total	**103**	**270**	**373**

EXPECTED:

	Hit	**Out**	**Total**
Ahead	$\frac{(140 * 103)}{373} = 38.7$	$\frac{(140 * 270)}{373} = 101.3$	**140**
Behind	$\frac{(104 * 103)}{373} = 28.7$	$\frac{(104 * 270)}{373} = 75.3$	**104**
Even	$\frac{(129 * 103)}{373} = 35.6$	$\frac{(129 * 270)}{373} = 93.4$	**129**
Total	**103**	**270**	**373**

TABLE 10.6: Observed and expected counts for the Alex Rodriguez batting outcome data in Example 2.

4 Other outcomes, for example, walks, are not part of these outcomes.

Step 3: We Set the Level of Significance at 0.05.

Step 4: Calculate Test Statistic

$$X^2 = \sum \frac{(O-E)^2}{E} = \frac{(44-38.7)^2}{38.7} + \frac{(96-101.3)^2}{101.3} + \frac{(22-28.7)^2}{28.7} + \frac{(82-75.3)^2}{75.3} + \frac{(37-35.6)^2}{35.6} + \frac{(92-93.4)^2}{93.4}$$

$$= 0.74 + 0.28 + 1.52 + 0.60 + 0.05 + 0.02 = 3.26$$

Substituting the data from Table 10.6 into our chi-square test statistic formula (if calculating by hand, remember to *square* each difference in the numerator),

Degrees of freedom (df) = (r - 1) x (c - 1) = (3 - 1) x (2 - 1) = 2

Step 5: Calculate p-value

With a chi-square test statistic of 3.26 and degrees of freedom of 2, we turn to the Chi-Square Table to find the p-value. Figure 10.3 provides a portion of the Chi-Square Table, with the row for degrees of freedom of two highlighted.

FIGURE 10.3: First four degrees of freedom chi-square values for various right-tail probabilities.

Chi-Square Table: Chi-Square (X^2) Values for Various Right Tail Probabilities

	Right Tail Probability						
df	0.250	0.100	0.050	0.025	0.010	0.005	0.001
1	1.32	2.71	3.84	5.02	6.63	7.88	10.83
2	2.77	4.61	5.99	7.38	9.21	10.60	13.82
3	4.11	6.25	7.81	9.35	11.34	12.84	16.27
4	5.39	7.78	9.49	11.14	13.28	14.86	18.47

We notice in the table that as we go from left to right, the chi-square values *increase* as the right-tail probabilities *decrease*. Going across that row of df = 2, we find that our chi-square value of 3.26 falls *between* chi-square values of 2.77 and 4.61, which have right-tail probabilities of 0.250 and 0.100, respectively. Therefore, our p-value will fall *between* 0.100 and 0.250.

Step 6: Conclusion

With our p-value range of 0.100 and 0.250 exceeding 0.05, then exact p-value would also exceed 0.05. We fail to reject the null hypothesis and conclude that there is not enough statistical evidence to support stating there is a relationship in the population for *pitch count* and *bat outcome* for Alex Rodriguez. Table 10.7 provides highlighted Minitab output for our test. Notice in this table how small each cell's chi-square contribution is. This illustrates that A-Rod's observed outcomes (how he actually performed in terms of *hits* or *outs*) in these pitch count situations were close to what was expected. His performance was steady - not to be interpreted as good or bad! - across the three levels of pitch count.

```
Observed counts listed first
Expected counts are printed below observed counts
Chi-Square contributions are printed below expected counts


           Hit     Out   Total
Ahead       44      96     140
         38.66  101.34
         0.738   0.281

Behind      22      82     104
         28.72   75.28
         1.572   0.600

Even        37      92     129
         35.62   93.38
         0.053   0.020

Total      103     270     373

Chi-Sq = 3.264, DF = 2, P-Value = 0.196
```

TABLE 10.7: Highlighted Minitab output for chi-square test of Alex Rodriguez data in Example 2.

Example 3: Is There a Relationship between "Scoring First" and "Winning" in the NHL Playoffs?

We wish to examine if scoring first in NHL playoff games is related to that team winning the game. This offers two levels, Yes or No, to the categorical variables of *Scored First* and *Won*. This example also provides a unique example of what we are testing when the sample sizes are equal for each category level. When one game is played, there are two outcomes: either a team wins or a team loses. Likewise, only one team can score first. To understand this concept more completely, consider the example when two games are played. There would be two winning teams, two losing teams, two teams that scored first, and two teams that did not score first. This would produce a table with each row and column total equaling two, and an overall total of four. In such situations, how would you *expect* the data to look if *independent*? That is, if there was no relationship between scoring first and winning, about how often would you expect the team scoring first two win? About half the time should make sense.[5] In other words, for example, if 50 games were played and the two variables were *independent*, we would expect about 25 times the team scoring first would win and about 25 times the team scoring first would lose. Thinking of it another way, we would expect the conditional percentages for each row to be 50%. In these situations of equal row and column totals for a two-by-two table, we can think of the chi-square test of independence as a test of whether the conditional percentages differ from 50%.

For our example, we will use results from the 2010–2011 NHL season, in which 89 playoff games took place (see www.nhl.com). With two levels to our two categorical variables of *Scored First* and *Won*, we would have a two-by-two table with a total of 89 in each row and column and 178 as the overall total. The observed data for our example is presented in Table 10.8.

	Won		
Scored First	No	Yes	**Total**
No	68	21	**89**
Yes	21	68	**89**
Total	**89**	**89**	**178**

TABLE 10.8: Observed counts for the 2010–2011 NHL playoffs in Example 3.

As stated previously, with equal sample sizes for rows and columns (89), we would expect about 45 times the team scoring first would win. However, from Table 10.5, this occurred 68 times, where the team scoring first also won. The test question becomes, do the observed counts provide evidence against the outcomes being evenly distributed if the two variables are independent? Using intuition, what should the expected table look like?

Assuming independence, we would expect each cell count to be around 45. Going through our expected count calculations, we get the counts given in Table 10.9.

	No	**Yes**	**Total**
No	$\frac{(89 * 89)}{178} = 44.5$	$\frac{(89 * 89)}{178} = 44.5$	**89**
Yes	$\frac{(89 * 89)}{178} = 44.5$	$\frac{(89 * 89)}{178} = 44.5$	**89**
Total	**89**	**89**	**178**

TABLE 10.9: Expected counts for the 2010–2011 NHL playoffs in Example 3 with Scored First in the rows.

As we thought, the expected counts are roughly 45 for each cell!

5 By expecting equal distribution of observations when row and column totals are equal, we would expect this distribution to be 1/2 for a two-by-two table, 1/3 for a three-by-three table, and so on.

Step 1: State Hypotheses

Ho: In the population, scoring first and winning are independent in NHL playoffs.

Ha: In the population, scoring first and winning are related (dependent) in NHL playoffs.

Step 2: Check Assumptions

We will assume the data from the 2010–2011 season represents a random sample of the population of all NHL playoff games. The two variables, *scored first* and *won*, are categorical. Since we have four cells, the assumption that at least 80% of the expected cell counts are at least five implies all cells in the expected table have counts of five or more. From Table 10.9, with all expected cell counts being 44.5, we have satisfied this assumption.

Step 3: We Set the Level of Significance at 0.05

Step 4: Calculate Test Statistic

Substituting the data from Tables 10.8 and 10.9 into our chi-square test statistic formula (if calculating by hand, remember to *square* each difference in the numerator),

$$X^2 = \sum \frac{(O-E)^2}{E} = \frac{(68-44.5)^2}{44.5} + \frac{(21-44.5)^2}{44.5} + \frac{(68-44.5)^2}{44.5} + \frac{(21-44.5)^2}{44.5}$$

$$= 12.41 + 12.41 + 12.41 + 12.41 = 49.64$$

Degrees of freedom (df) $= (r-1) \times (c-1) = (2-1) \times (2-1) = 1$

Notice that each cell's chi-square contribution is the same. This should occur when we have equal row and column totals, as the two rows are inverses of each other—as many "won scoring first" as "lost *not* scoring first," and vice versa.

Step 5: Calculate p-value

With a chi-square test statistic of 49.64 and degrees of freedom of 1, we turn to the Chi-Square Table to find the p-value. Figure 10.2 (repeated below) provides a portion of the Chi-Square Table, with the row for degrees of freedom of one highlighted.

FIGURE 10.2 (repeated): First four degrees of freedom chi-square values for various right-tail probabilities.

Chi-Square Table: Chi-Square (X^2) Values for Various Right Tail Probabilities

	Right Tail Probability						
df	0.250	0.100	0.050	0.025	0.010	0.005	0.001
1	1.32	2.71	3.84	5.02	6.63	7.88	10.83
2	2.77	4.61	5.99	7.38	9.21	10.60	13.82
3	4.11	6.25	7.81	9.35	11.34	12.84	16.27
4	5.39	7.78	9.49	11.14	13.28	14.86	18.47

We notice in the table that as we go from left to right, the chi-square values *increase* as the right-tail probabilities *decrease*. Going across that row of df = 1, we find that our chi-square value of 49.64 *exceeds* the last chi-square value of 10.83. This means that our p-value is *less than* 0.001.

Step 6: Conclusion

With our p-value being less than 0.001, the p-value would also be less than 0.05. We reject the null hypothesis and conclude that there is enough statistical evidence to support stating there is a relationship in the population of NHL playoffs between *scoring first* and *winning*. Table 10.10 provides highlighted Minitab output for our test. Based on our data, we would conclude that scoring first is related to winning playoff hockey games. Since we rejected the null hypothesis, we consider the chi-square contributions made by each cell to ascertain which levels of the variables might be supporting this relationship. Recall that a contribution value of zero indicates complete independence, as this would only occur when the observed and expected cell counts are equal.

```
Observed counts listed first
Expected counts are printed below observed counts
Chi-Square contributions are printed below expected counts

            No      Yes     Total

No          68      21       89
         44.50   44.50
         12.41   12.41

Yes         21      68       89
         44.50   44.50
         12.41   12.41

Total       89      89      178

Chi-Sq = 19.541, DF = 1, P-Value = 0.000
```

TABLE 10.10: Highlighted Minitab output for chi-square test of NHL data in Example 3 with Scored First in the rows.

As noted earlier, when we have the unique situation of equal row and column totals in a two-by-two table, the chi-square contributions will be identical. This makes the interpretation more straightforward, especially when we include the conditional percentages as given in Table 10.11.

	Won	
Scored First	No	Yes
No	76.4%	23.6%
Yes	23.6%	76.4%

TABLE 10.11: Conditional percentages for the 2010–2011 NHL playoffs in Example 3.

Assuming independence, we would have expected each conditional percentage to be 50%. With our test result being statistically significant, we are saying that these conditional percentages differ significantly from our expected percentages of 50% under independence.

10.4: Comparing Two Proportions and the Two-by-Two Table

As we went through our chi-square test of independence using two-by-two examples, one might wonder what the difference was between this test and our *Z*-test of two proportions in Chapter 8. The answer is nothing; the tests were identical. Whereas the *Z*-test considered was based on only the proportion of successes, the chi-square test incorporates successes and failures. The *Z*-test does offer the advantage of allowing us to conduct one-sided tests of one proportion being greater (or less than) the other proportion. The connection between the *Z*-test and chi-square test is the following:

When we have a test statistic that comes from the standard normal distribution (e.g., the Z-test statistic for comparing two proportions), then the square of this test statistic will follow a chi-square distribution with one degree of freedom.

If the *Z*-test is two-sided, then the p-value for that test and the chi-square test are the same, as is the conclusion for the two tests. To illustrate this, we recall Example 1 from Chapter 8, where we compared home win percentage between college and NFL teams. In that example, our alternative hypothesis was a difference in the two proportions. The sample data was:

College: Home team won 483 of 748 games

NFL: Home team won 145 of 256 games,

with test results of Z_{stat} of 2.26 and a 0.024 p-value.

If we apply a chi-square test of independence to this situation, using Minitab to help with the calculations, we get the following results shown in Table 10.12.

```
Observed counts listed first
Expected counts are printed below observed counts
Chi-Square contributions are printed below expected counts

              Won      Lost   Total
College       483       265     748
           467.87    280.13
            0.489     0.817

NFL           145       111     256
           160.13     95.87
            1.429     2.387

Total         628       376    1004

Chi-Sq = 5.122, DF = 1, P-Value = 0.024
```

TABLE 10.12: Chi-square test of home winning percentages for college and NFL teams—Example 1 from Chapter 8.

As you can see, the p-value is 0.024, as we attained in our *Z*-test. If we take the square root of the chi-square test statistic of 5.122, we arrive at 2.26, our *Z*-test statistic. Using either method, we would have rejected the null hypothesis and concluded a difference in winning percentages (*Z*-test) or a relationship between *Level of Football* and *Game Location*. From Table 10.12, we observed the college teams winning at home more than expected, while NFL teams lost at home more than expected. Turning this information into a one-sided *Z*-test, we would conclude that college teams are more likely to win at home than NFL teams—the same conclusion we drew in Chapter 8, since the home winning percentage for college teams (64.6%) was significantly greater than that in the NFL (56.6%).

10.5: Controlling for a Third Variable

Back in Section 2.1, we defined a confounding variable. For our purposes in this chapter, when our explanatory and outcome variables are related to a third variable—also an explanatory variable—this third variable is called a **confounding variable**. Not all third variables are classified as confounding. Take, for instance, our NHL play-offs example, where we analyzed scoring first and winning. A possible third variable is playing at home. That is, does playing at home make a difference in our analysis comparing scoring first and winning? When we add this variable, *game location*, into the data mix and then analyze the scoring first and winning based on where the game was played for the team scoring first (i.e., was the team that scored first also playing at home), we find no change in our result. When we split the data by where the game was played, the chi-square test statistic came to 24.75, with a 0.000 p-value. Therefore, we would conclude that scoring first is related to winning the game, regardless of whether the team scoring first was at home or on the road. The third variable, *game location*, would not be a confounding variable.

Sometimes, however, this third variable can have a remarkable effect of "flipping" the results of a study when the data is separated on this third variable. When the direction of a relationship changes by the inclusion of a third variable, this phenomenon is called **Simpson's paradox**. To illustrate Simpson's paradox, we break down the shooting data of LeBron James and that of James Harden of the Oklahoma City Thunder, the 2011–2012 winner of the NBA Sixth Man Award.

From Table 10.13, we can calculate the shooting percentage of both men. For LeBron, we get 621/1169 for 53.1%; for Harden, his shooting percentage is 309/629 = 49.1%. Based on these percentages, one might conclude that LeBron is a better shooter than Harden.

	Shot Result		
Player	Made	Miss	**Total**
LeBron	621	548	**1169**
Harden	309	320	**629**
Total	**930**	**868**	**1798**

TABLE 10.13: Shot results for all shots taken by LeBron James and James Harden during the 2011–2012 NBA season.

However, as we learned in Example 1 above, there can be a relationship between type of shot taken and shot result. If we broke down our Table 10.13 data into two- and three-point shots, would we still see LeBron as the better shooter? Table 10.14 shows separately each player's results for type of shot.

	LeBron			Harden		
	Shot Result			Shot Result		
Shot	Made	Miss	**Total**	Made	Miss	**Total**
Two pt	567	453	**1020**	195	142	**337**
Three pt	54	95	**149**	114	178	**292**

TABLE 10.14: Shot results broken down by shot type for LeBron James and James Harden during the 2011–2012 NBA season.

Once we break shots down into shot type, we see a reversal in the direction from the overall results. In Table 10.14, we can see that for both shot types, James Harden performs better than LeBron. For Harden, his shooting percentages were 57.9% for two-pointers and 39% for three-pointers, while for LeBron, these percentages were 55.6% for two-pointers and 36.2% for three-pointers. Regardless of the shot being for two or three points, James Harden made them at a greater percentage than LeBron James. Yet when we combined these results into an overall percentage, LeBron appeared as the better overall shooter. With type of shot acting as a third and confounding variable, when we do not control for it we have one result—LeBron is the better shooter. But when we do control for this variable, we have a conflicting result—Harden is the better shooter. As stated earlier, this reversal of results, once we control for a confounding variable, is known as Simpson's paradox.

However, do not confuse Simpson's paradox with a change of direction for fewer than all levels of the third variable. For instance, if in our example after controlling for *shot type*, LeBron was a better two-point shooter but Harden was the better three-point shooter, this would *not* have been a case of Simpson's paradox. The paradox relates to a change in direction in *all* levels of the third variable compared to the direction when this variable is not controlled.

10.6: Relative Risk

Although the discussion of **relative risk** appears in the topic of comparing two categorical variables, the concept involves the ratio of two proportions. The topic is most common in medical situations. Whenever one is reporting or reading about a relative risk, what is important to note is the **baseline risk**. As we will see, the relative risk is how likely one proportion is to occur compared to another proportion. Without knowing the baseline risk (i.e., the baseline proportion), making an educated interpretation of the relative risk proves problematic. To illustrate the concepts in bold font above, we will examine the current topic of concussions in the NFL; specifically, those that occur on kickoffs.

In an effort to reduce concussions, the NFL moved the spot of kickoffs up from the 30- to the 35-yard line. According to Atlanta Falcons president Rich McKay, this resulted in a 40% reduction in the number of concussions on kickoffs from 2010 to 2011 (see www.nfl.com). By reviewing game logs and aggregate season data at www.espn.go.com/nfl, we can determine the kickoff return data for the 2010 and 2011 NFL seasons. This data is given in Table 10.15.

	Kick Offs (KO)		
Year	KO Ret.	Total KO	Pct. Ret.
2010	2033	2495	83%
2011	1375	2449	55%

TABLE 10.15: NFL kickoff return data for 2010 and 2011 seasons.

As Table 10.15 demonstrates, the moving of the kickoff up five yards certainly had an effect in reducing the percentage of kicks that were returned. Assuming that a player is more likely to get a concussion on a returned kick compared to a non-returned kick (i.e., a touchback), then this move by the NFL should have had a positive impact on reducing the number of concussions suffered during kickoffs. And according to Rich McKay it did: a 40% reduction. However, Mr. McKay does not elaborate on how *many* concussions were experienced on kickoffs in either year. Without this information, the exact appreciation for this 40% reduction is proved difficult.

Consider these two hypothetical scenarios for the number of concussions sustained each year:

Scenario One: In 2010 there were 50 concussions on returned kicks, while in 2011 there were 30.

Scenario Two: In 2010 there were 5 concussions on returned kicks, while in 2011 there were 3.

In Table 10.16, we organize these two scenarios and include the risk (proportion) of concussion for each.

		Scenario One		Scenario Two	
Year	Returns	Concuss	Risk	Concuss	Risk
2010	2033	50	2.5%	5	0.25%
2011	1375	30	2.2%	3	0.22%

TABLE 10.16: NFL kickoff return data for 2010 and 2011 seasons along with two *hypothetical* scenarios.

In both scenarios, the *reduction* in number of concussions from 2010 to 2011 is 40%, as reported. But what are the relative risk and baseline risk, and what do they mean?

A relative risk is the ratio of two risks:

$$Relative\ Risk = \frac{Risk\ Group\ 1}{Risk\ Group\ 2}$$

How a relative risk is computed depends on which group's risk is being *compared to*. This group defines "Group 2." The "compared to" risk goes on the *bottom* of the relative risk. With the study comparing concussions from 2011 *to* 2010, the relative risk will use 2011 data as the numerator (Group 1) and 2010 data as the denominator in our risk (Group 2). This denominator risk, the Risk of Group 2, is referred to as the *baseline risk*.

From either scenario in Table 10.16, the relative risk is the same: 0.88, as found by:

$$Scenario\ One:\ Relative\ Risk = \frac{Risk\ 2011}{Risk\ 2010} = \frac{0.22}{0.25} = 0.88$$

$$Scenario\ Two:\ Relative\ Risk = \frac{Risk\ 2011}{Risk\ 2010} = \frac{0.022}{0.025} = 0.88$$

The interpretation of this relative risk is: *In 2011 the risk (proportion) of NFL players getting a concussion on a kickoff return was 0.88 times the risk (proportion) of NFL players getting a concussion on kickoff returns in 2010.*

If the study wanted to compare 2010 to 2011, the relative risk would be 1.14 (from 0.025 divided by 0.022), meaning risk of concussion on kickoff return in 2010 was 1.14 times the risk of getting a concussion on kickoff return in 2011.

Question: What would be the relative risk if there were *no change* in risk from one year to the next? The answer is one. If there were not a change in risks, then both risks would be the same. With the ratio of the two same numbers being one, the relative risk for two identical risks is also one. Therefore, risks further away from one (in either direction) indicate a stronger association. Let us consider these baseline risks.

In *Scenario One*, the baseline risk is 2.5%, while in *Scenario Two*, this is 0.25%. For the first scenario, this can be interpreted that for 100 kickoff returns there are 2.5 concussions, while for the second scenario there are 0.25 concussions—less than one!—per 100 kickoff returns. In both cases, the chances of concussion are small, but this is especially true in the case of *Scenario Two*. By knowing the baseline risk, one can make a better judgment on exactly what effect is occurring in a relative risk. If the baseline risk is extremely unlikely, then increasing (decreasing) the chances may have little meaning. For example, if relative risk of 0.88 is applied to *Scenario Two*, we would be saying that in 2011, for every 100 kickoff returns, there were 0.22 concussions. Not much difference from the 0.25 per 100 for 2010!

Of course, when dealing with an injury such as concussion that can often result in severe complications—both present and future—one should not take lightly any change in rules that improve player safety. However, in order to get a full understanding of the impact of such changes, one should provide both group risks, especially the baseline risk. This relates to our discussion in Section 7.5 about practical versus statistical significance. The data may (or may not) be statistically significant—we cannot determine this without all the information, but the NFL leaders may find the difference to be of practical importance.

Expressions and Formulas

1. Chi-square test of independence hypotheses

   Ho: In the population, the two variables are independent

   Ha: In the population, the two variables are dependent (associated)

2. Calculating expected counts (two-by-two table)

$$\textit{Expected Counts of each cell for row } (i) \textit{ and column } (j) = \frac{(\textit{Row Total}_i) x (\textit{Column Total}_j)}{\textit{Overall Total}}$$

3. X^2 test statistic of chi-square test of independence

$$X^2 = \sum \frac{(\textit{Observed Count} - \textit{Expected Count})^2}{\textit{Expected Count}} = \sum \frac{(O-E)^2}{E}$$

with df = $(r-1) \times (c-1)$, where r is the number of table rows (not including Total) and c is the number of table columns (not including Total)

4. Relative Risk is found by:

$$\textit{Relative Risk} = \frac{\textit{Risk Group 1}}{\textit{Risk Group 2}}$$

where *Risk Group 2* is the risk for the group being compared to.

Comparing Two Quantitative Variables—Correlation and Regression

11

In Chapter 10, our interest was in finding an association between two *categorical* variables. However, what if we have two variables that are *quantitative*?[1] For example, what if our research interests are in seeing if there is a relationship between player height and player weight? Or possibly we wish to examine for a relationship between a pitcher's earned run average (ERA) and the batting average (BA) of opposing hitters. In hockey, we may want to learn if team points are related to goals scored, or are they more strongly related to goals allowed, or possibly both. This would be a question of offense, defense, or a combination of the two being more important. In any case, these involve two quantitative variables. Often, with two quantitative variables, we are interested in exploring whether one variable can *explain*, or even cause, changes in another variable. As we have learned previously, we call the variable doing the explaining the **explanatory variable** and the other variable the **response variable**.

Once a relationship between two quantitative variables is established, a follow-up question could be, "What type of relationship is it?" The sports world is filled with applications where one or more variables are used in some function in an effort to estimate or predict some outcome. For instance, general managers in the various professional leagues will use a player's high school or college statistics to predict future success; agents will use statistics to estimate a player's financial value; team statistics are used to predict which team will win a particular game—or series of games—such as during March Madness; and the list goes on.

One particular example is the rule of 26-27-60 presented by John Lopez of *Sports Illustrated*. If a college quarterback scored at least 26 on the Wonderlic Test, started at least 27 games in college, and completed least 60% of his passes, then he has a good chance to succeed in the NFL. Several questions one might ask are:

1. Of all the data on players, how did Lopez arrive at these three specific variables, as well as these minimum scores, being significant to predicting success?
2. How successful is his model in predicting success? Is it right all the time? 75% of the time?
3. Could there be other variables better related to NFL success than the variables Lopez chose?
4. What is the relationship between these variables and NFL success?

---

1 In the event you have two variables, one categorical and one quantitative, you can consider side-by-side box plots for graphic representation or recall the test of two independent means!

5. Do these variables form an equation, and if so, how does one use it?

6. How is success being defined?

Here in Chapter 11, we will focus on one particular relationship: the linear one. The methods we discuss will relate to how we determine if two quantitative variables are linearly related, and if so, how does one go about finding the equation of a line that would best fit the data in order to predict some outcome. This chapter will be more technical than previous ones, speaking to an appreciation for the development of statistical software.

11.1: Examining a Relationship between Two Quantitative Variables

To begin, we return to the heights of the 13 players on the Duke men's basketball team for 2011–2012. Along with each player's height, we now include his weight. Our question is whether we can show that player height is *linearly related* to player weight. This data is presented in Table 11.1, in increasing order of players' heights.

Height	**Weight**
72	175
73	195
74	180
76	200
76	200
79	205
79	235
80	220
81	240
82	235
82	245
83	230
83	225

TABLE 11.1: Heights and weights for players on the 2011–2012 Duke men's basketball team.

As we proceed through our discussion, this data set offers an opportunity to illuminate several key thoughts to keep in mind.

One, notice how, in general, as *height* increases so, too, does *weight*. This represents a *positive* relationship between x, height, and y, weight. This pattern of "as X increases, so does Y" is indicative of a positive relationship. However, *negative* relationships can also occur. These will be evident when "as X increases, Y decreases." As an example of a negative relationship, consider what happens to an NFL player's forty-yard dash time as his weight increases. Here, weight would be x, influencing forty-yard time, y.

Two, not all Y values are the same for equal X values. In our Duke data, we see that the two players with a height of 72 inches both weigh 200 pounds, but the two players of height 79, 82, and 83 inches all have different weights; 205 and 235 pounds, 235 and 245 pounds, 230 and 225 pounds, respectively. What this tells us is, although there might be a linear relationship between height and weight, this relationship is not perfect. In other words, height might explain *some* of the reasoning behind why players have different weights, but it does not explain *all* of the reasoning. In the study of statistics, this fact will almost always be true—even if the data set is so large that you cannot see multiple *y* values for equivalent *x* values.

Finally, there are some players who weigh less than someone who is shorter. For example, one player who is 83 inches tall weighs 225 pounds, which is less than a player who is 79 inches tall and weighs 235 pounds. The former is 4 inches taller, yet 10 pounds lighter. Such observations may reveal themselves as **outliers**, also called **extreme observations**. We might recall this term from Chapter 1.

A typical data set will not be ordered (although one can do so) and will often be much larger in terms of the number of observations. This is why graphing is so important. As with many areas in statistics, graphs play an important role here as well. They offer insight into the shape of our data, allow us to compare groups of data, and so on. Whenever we have two quantitative variables, we should begin by plotting the data. The proper graph for such situations is a **scatterplot**. A scatterplot is a two-dimensional graph displaying the (x, y) pairs of observations. We plot on the vertical axis the *y* values of our response variable, with the corresponding *x* values of our explanatory variable along the horizontal axis. Figure 11.1 provides a scatterplot of our Duke men's basketball data.

FIGURE 11.1: Scatterplot of heights and weights for players on the 2011–2012 Duke men's basketball team.

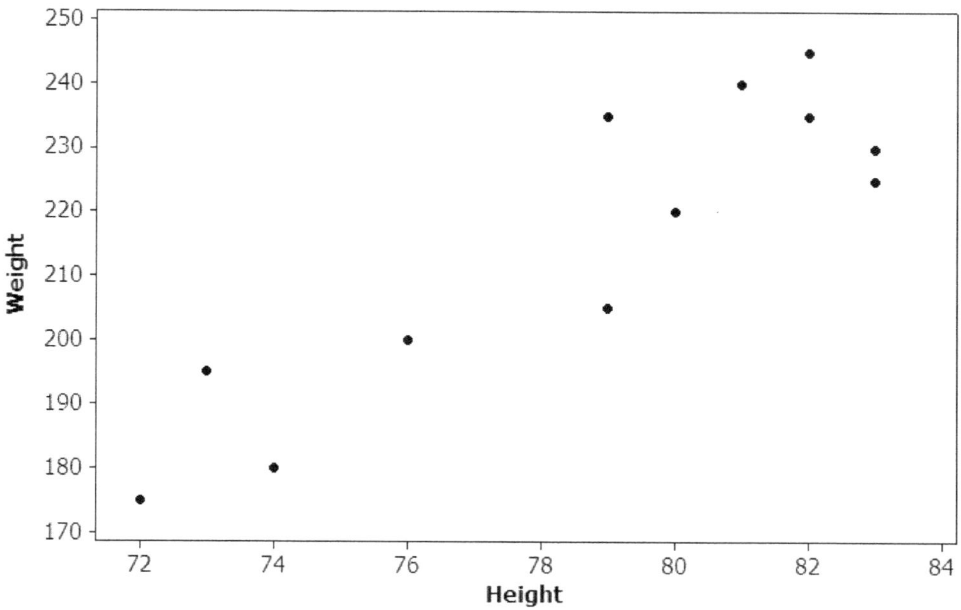

The scatterplot supports our earlier observation from the data of Table 11.1. We have a positive relationship between *height* and *weight*, and for some equal height values we have different weights. As you inspect this graph, picture a line drawn through these plotted points that might "fit" this plot in such a way that we could use the equation of the line to predict weight from height. This allows us to consider not just any relationship between the two variables, but more specifically a *linear* relationship. However, with the variation we have in weights for players of equal heights, any line we establish will include some error in our prediction. For instance, if a player had a height of 82 inches, the equation could not predict a weight of 235 *and* 245. This error would arise by comparing what weight we *observed* to the weight we *predicted*. We will learn more about these errors when we study how to find a best-fit equation for two quantitative variables in Section 11.2.

With a graphing mechanism in place, we need to put some numerical perspective on this linear relationship. A statistic called the **correlation**, denoted by *r*, is used to measure the strength and direction of a linear relationship between two quantitative variables. The correlation will often be calculated using software, but we provide the equation here, as all of the pieces used in the calculation are familiar.

$$r = \frac{1}{n-1}\sum\left(\frac{x_i - \bar{x}}{S_x}\right)\left(\frac{y_i - \bar{y}}{S_y}\right)$$

where n is the number of (x, y) paired data points, $\bar{x}$ and $\bar{y}$ are sample means, and S_x and S_y are sample standard deviations of x and y, respectively. The correlation is characterized by the following properties:

- A positive r value indicates a positive linear relationship, while a negative r value indicates a negative linear relationship.
- r is constrained by -1 and 1. That is, $-1 \leq r \leq +1$, with a value of 0 indicating no linear relationship. Therefore, as r approaches 0, the linear relationship gets weaker; as correlation r moves away from 0, the linear relationship gets stronger. A perfect linear relationship will occur when r is -1 (perfect negative linear relationship) or r is +1 (perfect positive linear relationship). Keep in mind that the *sign* of the correlation provides the *direction* of the linear relationship—positive or negative—while the actual *value* offers the *strength* of the linear relationship.
- There are no units of measurement connected with r. Looking back on our formula for r, the equation uses the standardized values of x and y, thus removing the units of measurement.
- r only measures the strength and direction of a *linear* relationship. The data might be related in another way—say, curved—resulting in a correlation value close to 0.
- Outliers can affect r.
- r will have the same sign as the slope of the best-fit line for the data, and vice versa.

From our Duke men's basketball data, where x is height and y is height, we have the summary measures: $n = 13$, $\bar{x} = 78.46$, $S_x = 3.86$, $\bar{y} = 214.23$, and $S_y = 23.17$. Table 11.2 shows the hand and Minitab calculations for the correlation. The slight difference is due to rounding error in our "by-hand" process.

x_i	y_i	$\frac{x_i - \bar{x}}{s_x}$	$\frac{y_i - \bar{y}}{s_y}$	$\left(\frac{x_i - \bar{x}}{s_x}\right)\left(\frac{y_i - \bar{y}}{s_y}\right)$
72	175	-1.67	-1.69	2.82
73	195	-1.41	-0.83	1.17
74	180	-1.15	-1.47	1.69
76	200	-0.64	-0.61	0.39
76	200	-0.64	-0.61	0.39
79	205	0.13	-0.4	-0.05
79	235	0.13	0.9	0.12
80	220	0.38	0.25	0.1
81	240	0.64	1.11	0.71
82	235	0.9	0.9	0.81
82	245	0.9	1.33	1.2
83	230	1.15	0.68	0.78
83	225	1.15	0.47	0.54
$\bar{x}$ = 78.46	$\bar{y}$ = 214.23			$\sum$ = 10.67
S_x = 3.86	S_y = 23.17			$r = \frac{10.67}{13-1} = 0.89$
Minitab calculation of r = 0.898				

TABLE 11.2: Hand and Minitab calculations of correlation between heights and weights for players on the 2011–2012 Duke men's basketball team.

This correlation value suggests a strong, positive linear relationship between height and weight for Duke men's basketball. Notice that our correlation was consistent with the properties of r:

- From the scatterplot in Figure 11.1, we knew the relationship was positive; as height increased so, too, did weight. This was reflected with $r > 0$.
- r fell within the range of $-1 \leq r \leq +1$.
- No units of measurement are associated with r. We do not list some "pounds per inch" or anything similar with our correlation of r. We just offer the *value* of r.
- With our scatterplot in Figure 11.1 suggesting a linear relationship between height and weight, r is a valid measure of the direction and strength of this linear relationship.
- The variables x height and y weight are both quantitative.

11.2: The Least Squares Regression Line

When the scatterplot and correlation indicate a linear relationship, a straight line is used to describe this relationship. However, we can draw many lines through the data points. Figure 11.2 illustrates just four possible lines we could draw.

FIGURE 11.2: Example of four possible lines drawn to fit scatterplot of heights and weights for players on the 2011–2012 Duke men's basketball team.

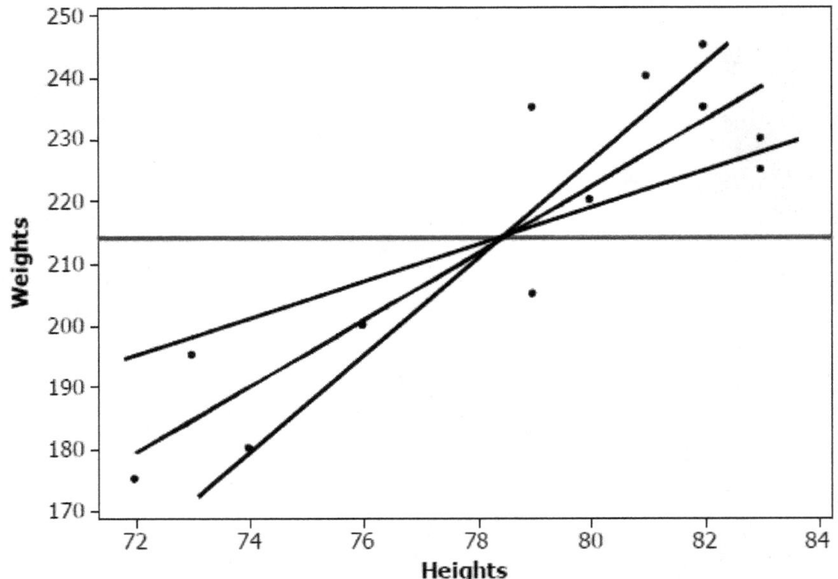

The horizontal line in Figure 11.2 represents the mean of y, Weights. Being horizontal, the line has zero slope, and therefore would represent a situation where the two variables did not have a linear relationship: our "best guess" for predicting a player's weight would be to simply use the average weight of the sample. However, in our example, we know there is a positive, linear relationship between *height* and *weight*. Therefore, either of the other three lines should better fit the data than the horizontal line; but which of these three is best and is that one the best fit line overall? How would we find such a best-fit line?

In statistics, the best fit line for two quantitative variables is called a **regression line**.[2] We want to use a straight line to depict how a response variable, y, tends to change as the explanatory variable x changes. This straight line is represented by the linear equation,[3]

$$\hat{y} = a + bx$$

The symbol $\hat{y}$ (pronounced "y hat") represents the *fitted* (also called *predicted*) values of y that the line would make for the observed x values. Because of this, the regression equation is sometimes referred

2 Why the term *regression*? The concept derives from research conducted by Sir Francis Galton in the 1800s, where he demonstrated that children of tall parents tend to *regress* toward the mean.

3 For those who have studied algebra, this equation may look familiar to y = mx + b, where "m" denotes slope and "b" denotes the y-intercept. Also, other statistics texts or instructors may use where "b_0" denotes the y-intercept and "b_1" the slope. Consider these equivalent, as in each case the constant is the y-intercept and the coefficient of x is the slope.

to as the **prediction equation**. Also, since the variable x is used to *predict*, we often refer to x as a **predictor variable**. We can use the terms *predictor* and *explanatory* interchangeably in regression. In the equation, "a" represents the ***y*-intercept** and "b" represents the **slope**. The **interpretation of slope** is:

For a unit change in x, the predicted value of y will increase (or decrease) in units by the value of the slope.

With this equation, how do we go about getting the y-intercept and slope for the *best* regression line? For starters, with one use of the regression line for *predicting* an outcome—and we illustratec earlier in Section 11.1 that this prediction will not be perfect; there will be some error—then one feature of the best line should be that it provides the best predictions. In other words, the best-fit line should result in the least **prediction error**. Of all possible lines from which we could choose, the best-fit line should make predictions of y closest to the observed values of y better than all the other lines. To accomplish this, a method called **ordinary least squares** is used. This method produces the y-intercept and slope for a line that results in the sum of the squared errors, also called **residuals**, being a minimum. The solution for this turns into a calculus problem, but fortunately with a unique solution assuring us of arriving at y-intercept and slope for the overall best-fit line. These solutions are,

$$b = r\left(\frac{S_y}{S_x}\right) \text{ and } a = \bar{y} - b\bar{x}$$

with r being the correlation between x and y, S_y and S_x representing the standard deviation of y and x, respectively, and $\bar{y}$ and $\bar{x}$ the mean of y and x, respectively. Recalling from Section 11.1 with our discussion on the heights and weights of the Duke men's basketball team, we had $\bar{x} = 78.46$, $S_x = 3.89$, $\bar{y} = 214.23$, $S_y = 23.17$, and $r = 0.898$ (from Minitab in Table 11.2). Plugging this data into the formulas for slope and y-intercept, we get,

$$b = r\left(\frac{S_y}{S_x}\right) = 0.898\left(\frac{23.17}{3.86}\right) = 5.39 \text{ and}$$

$$a = \bar{y} - b\bar{x} = 214.23 - 5.39(78.46) = 214.23 - 422.90 = -208.6$$

Putting these into our line, we get the best-fit regression equation of,

$$\hat{y} = a + bx = -208.6 + 5.39x$$

Applying our interpretation of the slope to this example, we would say, *for an increase in a player's height of one inch, a player's predicted weight will increase by 5.39 pounds.*

To get the *predicted* or *fitted* weights for our Duke players, we would plug their x height values into the regression equation. The difference, then, between the players' observed weights and predicted weights represents the *prediction errors*, or **residuals**. As an example, consider the (height, weight) pairs (73, 195) and (74, 180), which are the second and third listed observations in Table 11.1. Plugging the height values into our regression equation, we get predicted weights,

$x = 73$: $\hat{y} = -208.6 + 5.39x = -208.6 + 5.39(73) = 184.87$

$x = 74$: $\hat{y} = -208.6 + 5.39x = -208.6 + 5.39(74) = 190.26$

Note two aspects of the predicted weights:

- That the difference in heights was one inch (i.e., one unit), resulting in a change of predicted weights of 5.39 pounds, the value of the slope.
- There is error in both predictions. For a player with a height of 73 inches, his predicted weight was 184.87 pounds, while his actual (or observed) weight was 195 pounds. His *residual* is 195 - 184.87 = 10.13 pounds. Likewise, the player with a height of 74 inches has a predicted weight of 190.26 pounds, with an actual weight of 180 pounds. His *residual* is 180 - 190.26 = -10.26 pounds. The first player weighs *more* than predicted, while the second player weighs *less* than predicted.

Figure 11.3 shows the scatterplot with the fitted regression line. Also included are lines indicating the residuals for heights of 73 and 74 inches. These are the straight lines drawn from the points to the equation line.

FIGURE 11.3: Regression line fit to scatterplot of heights and weights for players on the 2011–2012 Duke men's basketball team. Also shown are the residual lines for players with heights of 73 and 74 inches.

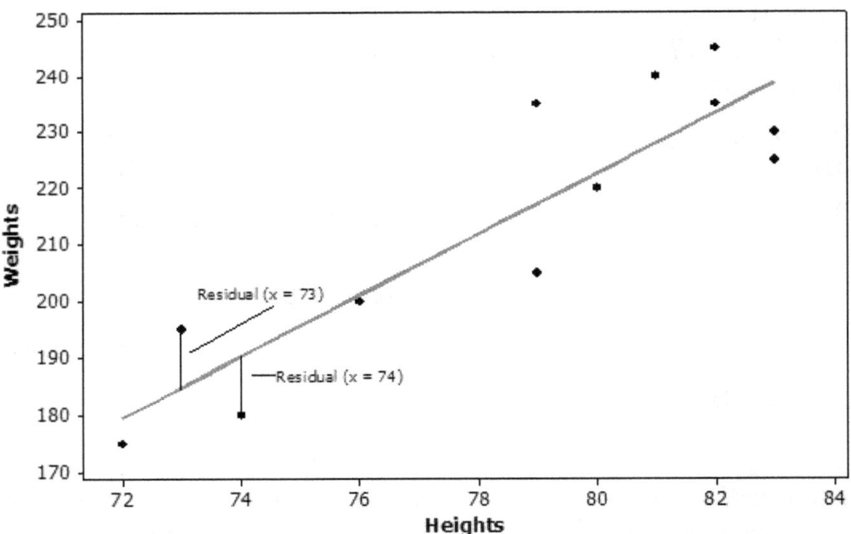

From Figure 11.3, we notice that points above the line have a positive residual and points below the line have a negative residual. Any point on the line would have zero residual. As discussed previously, the idea behind fitting a line to a set of data is to select one that produces the y-intercept and slope for a line where sum of the squared error is a minimum. This translates in the formula,

$$\textit{Error sum of squares} = SSE = \sum (error)^2 = \sum (Y_i - \hat{Y}_i)^2$$

where SSE denotes **sum of squared error**, Y_i is the observed *y* value for observation *i*, and $\hat{Y}_i$ is the predicted *y* value for observation *i*. The best-fit line is the one with the minimum sum of squared error, hence the name given to this technique to find the regression line: *least squares method*. The regression line will have the properties:

- The sum of the residuals will be zero.
- The regression will pass through the point of $(\bar{x},\bar{y})$.

With the sum of the residuals equaling zero, any "too high" predictions will be counterbalanced by "too low" predictions. Having the line pass through the point of the means for *x* and *y* indicates the regression line goes through the center of our data.

Evaluating *x* as a Predictor of *y*

Once we have determined the correlation and regression line for two quantitative variables, we would want to know how well the line works in predicting *y*. Earlier, we mentioned that if we only had observations of *y*, the best option to predict *y* would be to simply use the mean. For instance, if we just had the weights of the Duke players, the best prediction of weight would be the mean of the weights. By including a second variable linearly related to *y*, we would expect our prediction of *y* to improve; our prediction would have less error, thus be more accurate. The inclusion of height was one such variable. We demonstrated that height was linearly related to weight and developed a regression equation, but how can we determine if using this line with height would be any better than using only the mean of weight? By understanding there is variability in *y*—that not all of the players have the same weight—then a good predictor variable should be one that helps *explain* this variation in *y*. Thus the name *explanatory* variable. If we ignored *x* and went just with the mean of *y*, we would have the following sum of squared errors:

$$SST = \sum (Y_i - \bar{Y})^2$$

where SST stands for **sum of squares total** or **total sum of squares**.[4] When we add an explanatory variable to our prediction of *y*, we have our sum of squared error (SSE) from above,

$$SSE = \sum (Y_i - \hat{Y}_i)^2$$

When our linear relationship between *x* and *y* is strong, the predictions are a better choice than the mean . As a result, our SSE is much smaller than SST, since our predictions are closer to the observed *y* values than is the mean. By creating a ratio of these two error summaries, the result is a measure of the proportion of reduction in error. We call this the **coefficient of determination**, denoted by R-squared (R^2). The formal equation is,

4 Recall we introduced this term in Chapter 9 when we learned about Analysis of Variance (ANOVA).

$$R^2 = \frac{\sum(Y_i - \bar{Y})^2 - \sum(Y_i - \hat{Y}_i)^2}{\sum(Y_i - \bar{Y})^2} = \frac{SST - SSE}{SST} = 1 - \frac{SSE}{SST}$$

Typical R^2 is reported as a percentage by multiplying the above result by 100%. The interpretation of R^2 becomes *the proportion of variation in y that is explained by x*. The higher the value of R^2, the better x is performing to predict y compared to the mean of y. The properties of R^2 are,

- As a percentage, $0\% \leq R^2 \leq 100\%$.
- As R^2 approaches 0, the worse the regression equation works in predicting y; as R^2 approaches 100%, the better the regression equation works in predicting y.
- If R^2 is 0, then x is not linearly related to y. From the R^2 formula, this will occur when SST = SSE; the regression line is no better at helping to predict y than is the mean of y.
- If R^2 is 100%, then x has a perfect linear relationship with y. From the R^2 formula, this will occur when SSE is 0; the regression equation perfectly predicts y; there is no error in the prediction of y when using the regression equation.
- When we have two quantitative variables, one serving as the response variable and the other as the explanatory variable, **the value of R^2 can be found by squaring the correlation (r)**.

Applying property 5 to the Duke basketball example, where $r = 0.898$, we have,

$$R^2 = (0.898)^2 = 0.806 \text{ or } 80.6\%$$

Interpreting this value of R^2, we would say, *80.6% of the variation in weights for the Duke men's basketball team can be explained by height.* With the maximum possible value being 100%—and extremely unlikely—height is doing a very good job in predicting weight compared to using only the mean weight.

What Is the Regression Line Actually Predicting?

Throughout our discussion, we have referenced x as predicting some y outcome. However, we have also shown that not all y values will be the same for some given value of x. There is variability not only in all the y values, but also within y values for the same x value. As an example, we pointed out three instances in our Duke data where players of the same height had differing weights. If we consider this data to be a sample of *all* Duke men's basketball players, the predicted value of y actually represents the *estimated mean* value of y for given levels of x. When we predicted a of 190.26 pounds for a player 74 inches tall, we can think of this as the estimated mean weight for *all* Duke players that are 74 inches.

11.3: Inference about the Regression Line

So far, we have discussed some summary measures in defining the relationship between two quantitative variables, mainly correlation (*r*) and the coefficient of determination (R^2). We introduced methods for finding the best-fit line called the regression equation in order to use *x* to predict *y*. However, we have not discussed how to determine if the explanatory variable *x* is a statistically significant linear predictor of *y*.

As we have learned so far, one purpose for using an explanatory variable *x* is to provide a better prediction of *y* compared to the mean of *y*. Earlier, we referenced in Figure 11.2 that the line representing the mean of *y* would be a horizontal line, which has zero slope. It should stand to reason then that the use of regression methods would be considered better at predicting *y* if the slope of the regression line was significantly *different* from zero (recall we are saying "different," since the slope could be positive *or* negative). We will quickly learn the advantages of using software in regression, as many of the necessary calculations are quite muddled.

Up to this point in the example using the 2011–2012 Duke men's basketball team as our data set, we did not differentiate between this being sample data or population data. With the discussion moving on to inference in regression, we need to consider the data as a sample taken from some larger population. By doing so, we can establish a **population regression equation**, whereby the *y*-intercept (a) and slope (b) of our regression equation can be considered *point estimates* of the population *y*-intercept and population slope. This population regression equation can be written as[5],

$$\mu_y = \alpha + \beta X$$

where μ_y denotes the population mean of *y*; α is the population *y*-intercept estimated by *a*; and β is the population slope estimated by *b*.

Step 1: State Hypotheses

As always, we have two competing hypotheses, both of which use parameter notation. The expression "b" in our regression equation denotes the sample slope, as the value is estimated from the sample data. This sample slope is an estimate of the population slope, denoted β.

Ho: $\beta = 0$
Ha: $\beta \neq 0$

5 Another common set of terminology is to use B_0 and B_1 or β_0 and β_1 to represent the population y-intercept and population slope, respectively.

Step 2: Check Assumptions

- The data represent independent random samples
- There is a linear relationship between x and y
- In the population, the y values follow a normal distribution at each value of x
- In the population, the standard deviation (or variance) of the y values is the same at each value of x.

The first assumption depends on how the sample was taken. If taken randomly, we can assume independence. The remaining assumptions will be supported by scatterplots, probability plots, and plots of the residuals against x, this last graph is referred to as a **residual plot**. With a residual plot, we want to see a random "scatter" of residual values across the values of x. Think of this as taking a paintball gun and just randomly firing at a big, white wall. When finished, if randomly done, the points would show no discernible pattern.

Step 3: Set Level of Significance (per prior tests in this book, we have fixed this at 0.05)

Step 4: Calculate Test Statistic

Although we are going to present the formula for the test statistic, the values needed to compute the test statistic will come from the software. The software will also include the value of the test statistics and resulting p-value for our hypothesis test listed above. However, it is necessary to know where to find these values in the output. The test statistic for the slope comes from the T-distribution, and as we recall, this distribution is associated with a degrees of freedom measure. The test statistic is found by,

$$t = \frac{b - 0}{se(b)}$$

where the slope b and standard error of b will come from the software. The degrees of freedom for this test are df = n -2, where n is our sample size.

Step 5: Calculate p-value

With our alternative hypothesis being a "not equal" test and our test statistic coming from the T-distribution, the method of finding the p-value will follow the same process as that in Chapter 7 for a two-sided alternative hypothesis. However, the p-value will be furnished by software. Our main focus will be on correctly interpreting this output.

Step 6: Conclusion

As with all of our tests to this point, our decision is to *reject* the null hypothesis when our p-value is less than our 0.05 level of significance. When we reject Ho in a regression test of the slope, we conclude that there is enough evidence to claim that the *population* slope is significantly different from zero. Furthermore, we can state that x is a statistical linear predictor of y.

Example 1: Duke Men's Basketball (continued)

We will use the Duke men's basketball team to demonstrate how to conduct a regression analysis. What we want to show is that *height* is a significant linear predictor of *weight* for the Duke team.

Step 1: State Hypotheses

Ho: $\beta = 0$
Ha: $\beta \neq 0$

Step 2: Check Assumptions

A. Independence: Without stating that the 2011–2012 team was randomly selected from *all* Duke men's basketball teams, the sample is one of convenience and not randomness. Therefore, we could not assume independence. As a result, if we wanted to extend any inference results to the population of all Duke teams, we should do so with caution.

B. The scatterplot in Figure 11.1 revealed a linear relationship between *height* and *weight*. The figure is repeated below.

FIGURE 11.1 (repeated): Scatterplot of heights and weights for players on the 2011–2012 Duke men's basketball team in Example 1.

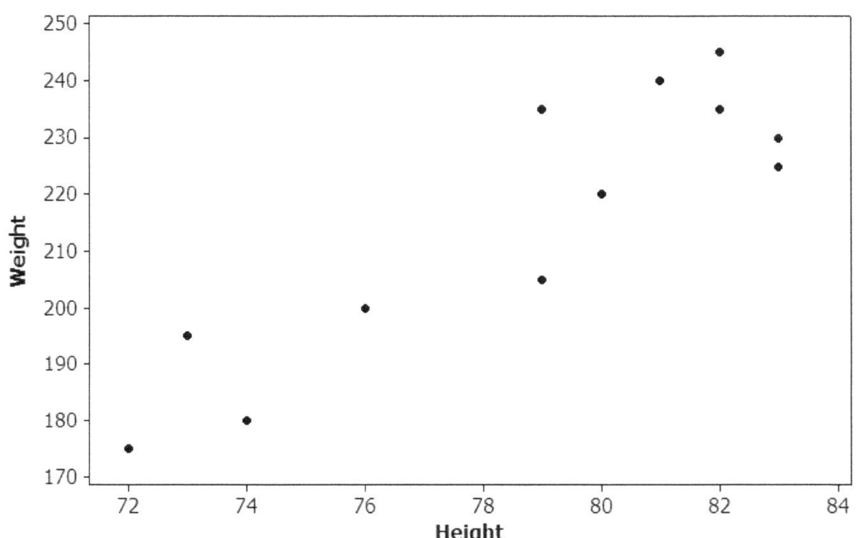

C. To check for normality, we can use a box plot or a probability plot of the residuals. A probability plot is provided in Figure 11.4, with no indications of a departure from normality. We base this decision on the p-value (0.421) for the test of normality being greater than 0.05.

FIGURE 11.4: Probability plot of residuals for heights and weights for players on the 2011–2012 Duke men's basketball team in Example 1.

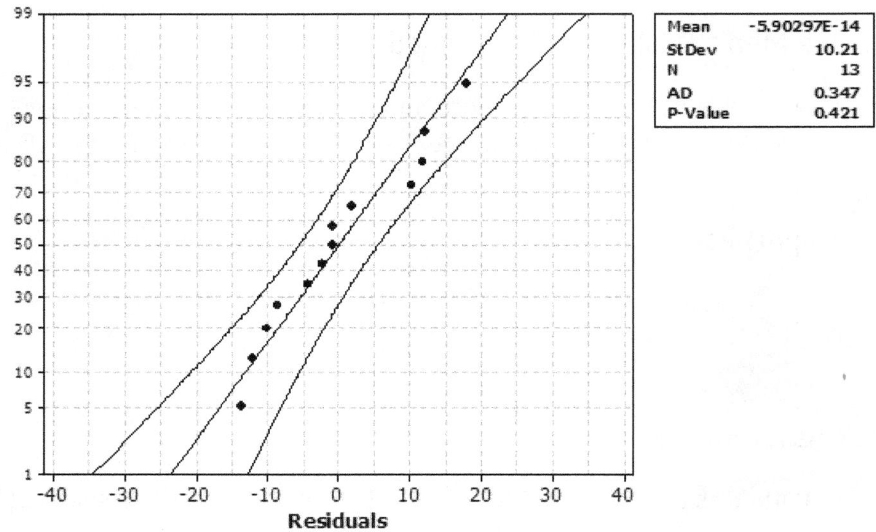

D. To check to see if our data provides evidence of equal standard deviations of y across all values of x, we plot the residuals against height. This plot is given in Figure 11.5.

FIGURE 11.5: Plot of residuals versus heights for players on the 2011–2012 Duke men's basketball team in Example 1.

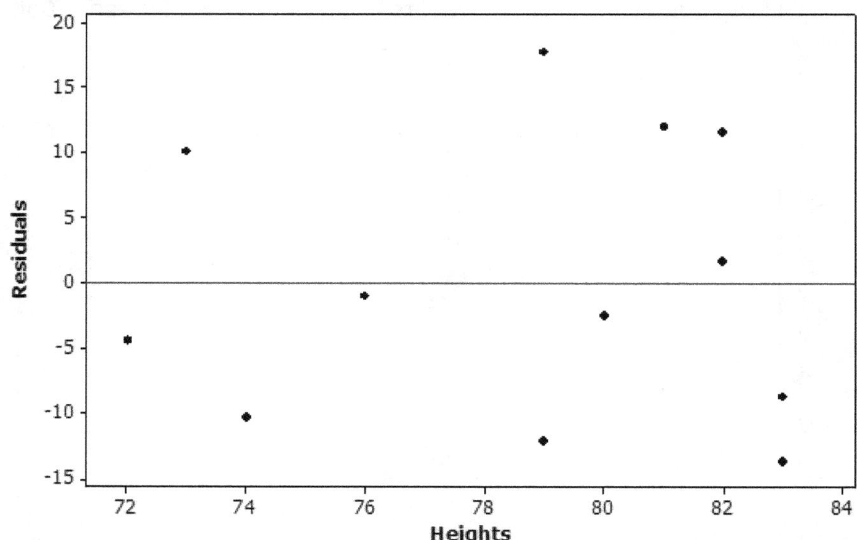

A plot of the residuals against the predictor variable should show a random pattern of residuals on both sides of 0. If a point lies far from the majority of points, it may be an outlier. Also, there should not be any recognizable patterns in the residual plot. Such patterns where the residuals fan out as x increases or where the residuals increase as x increases would be indicators of the standard deviations not being equal across all levels of x. This would lead to a conclusion that the assumption is not satisfied. In Figure 11.5, no pattern appears. The points shows a nice, even scatter around the zero-reference line. We can conclude the assumption of equal standard deviations for *weights* across all values of *height* is met.

Step 3: We Set the Level of Significance at 0.05

Step 4: Calculate Test Statistic

At this point, we will introduce output from Minitab to assist us in calculating the test statistic and p-value. Table 11.3 is annotated Minitab output.

```
The regression equation is Weights = - 208 + 5.38 Heights

              y-intercept
Predictor    \ Coef    SE Coef       T        P
Constant    -208.05     62.56    -3.33    0.007
Heights      5.3820    0.7965     6.76    0.000
               ↑          ↑         ↑        ↑
             Slope      se(b)    T test   p-value for test
                                 statistic of Ho: β = 0

R-sq = 80.6% ← coefficient of
               determination R²
```

TABLE 11.3: Annotated Minitab regression output for players on the 2011–2012 Duke men's basketball team in Example 1.

From the output under the heading "Coef" are the regression line y-intercept and slope, respectively. The software rounds these values in providing the final regression equation of,

$$\text{Weights} = -208 + 5.38 \text{ Heights}$$

We matched closely in Section 11.2 with a hand calculation of,

$$\hat{y} = -208.6 + 5.39x$$

From the output, we can reconstruct the t-test statistic for the slope in Table 11.3 by,

$$t = \frac{b - 0}{se(b)} = \frac{5.3820 - 0}{0.7965} = 6.76$$

With a sample size of n = 13, the degrees of freedom are n - 2 = 13 - 2 = 11.

Step 5: Calculate p-value

From the Minitab output in Table 11.3, we find a p-value of 0.000 for the test of Ho: $\beta = 0$. We can also find this using the T-Table, a portion of which is provided in Figure 11.6.

FIGURE 11.6: Portion of T-Table giving various *t* values with right-tail probabilities.

T-Table: t Distribution Confidence Interval and Critical Values

	Confidence Level					
	80%	90%	95%	98%	99%	99.8%
	Right Tail Probability					
df	$t_{0.10}$	$t_{0.05}$	$t_{0.025}$	$t_{0.01}$	$t_{0.005}$	$t_{0.001}$
1	3.078	6.314	12.706	31.821	63.657	318.289
2	1.886	2.920	4.303	6.965	9.925	22.328
.	....	....	....	....	....	....
.	....	....	....	....	....	....
11	1.363	1.796	2.201	2.718	3.106	4.025

Comparing the absolute value of the test statistic 6.76 to the *t* values in Figure 11.6 for df = 11, we see that 6.76 *exceeds* the largest *t* value of 4.025, corresponding to a 0.001 right-tail probability. Since our alternative hypothesis Ha is "not equal," we double this right-tail probability to conclude that the p-value would be less than 0.002 for our test of Ho: $\beta = 0$. This agrees with the Minitab output p-value of 0.000.

Step 6: Conclusion

With the p-value being less than 0.05, we reject the null hypothesis. We have enough statistical evidence to conclude that *height* is a significant linear predictor of *weight* for the men's basketball teams at Duke. The output also includes the coefficient of determination R-sq, or R^2, of 80.6%. The interpretation of this R^2 is that 80.6% of the variation in the weights of men basketball players at Duke is explained by their heights.

Example 2: Defense and Scoring in the NHL

In the NHL, a team gets 2 points for a win, 1 point for an overtime loss, and 0 points for a loss during regulation (i.e., no overtime). Teams are then ranked from highest point to lowest point totals, with the top teams (8) in each of two conferences making the playoffs. As with many sports, one may argue over what is more important to a team's success: offense or defense. Naturally, being proficient in both would provide the best scenario, but that is not always possible. We will examine whether defense leads to higher point totals. As a measure of defense, we use *goals against*, which represent the number of goals allowed by the team during the season. This serves as the explanatory variable *x*. The response variable will be the *points* accumulated by the team. We will use the final regular season results from the 2011–2012 NHL season as a sample of all NHL seasons.[6] With the NHL composed of 30 teams, we have a sample size of n = 30.

Step 1: State Hypotheses

Ho: $\beta = 0$
Ha: $\beta \neq 0$

Step 2: Check Assumptions

A. Independence: If we randomly selected this season from all NHL seasons, we could assume independence. This would allow us to feel more comfortable in extending any inferences we make to the population of all NHL seasons.

B. The scatterplot in Figure 11.7 reveals a *negative* linear relationship between *goals against* and *points*. As a team allowed more goals, the team points accumulated decreases. This would seem logical, given that one would probably assume that if a team gave up more goals, their chances of winning would decrease, thereby lowering their point total. From Minitab software, we have a correlation $r = -0.685$, with the negative sign in agreement with the scatterplot.

6 This scoring system was adopted at the beginning of the 2005–2006 season. Since the sample must be representative of the population, claiming the sample is from the population of ***all*** NHL seasons would only include all seasons since the scoring changes were implemented.

FIGURE 11.7: Scatterplot of goals against and points for NHL teams during the 2011–2012 regular season in Example 2.

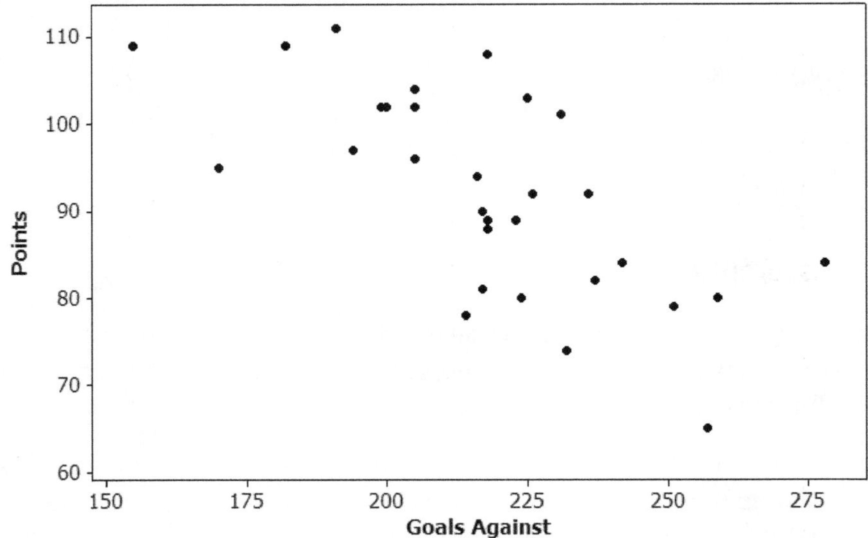

C. To check for normality, we can use a box plot, histogram, or a probability plot of the residuals. For this example, we will use a histogram, which is provided in Figure 11.8. From the graph, the shape appears approximately bell shaped, with no indications of a departure from normality. A probability plot of the residuals (not shown) resulted in a p-value of 0.524, which further supports the decision of the normal assumption being satisfied.

FIGURE 11.8: Probability plot of residuals for heights and weights for players on the 2011–2012 Duke men's basketball team in Example 1.

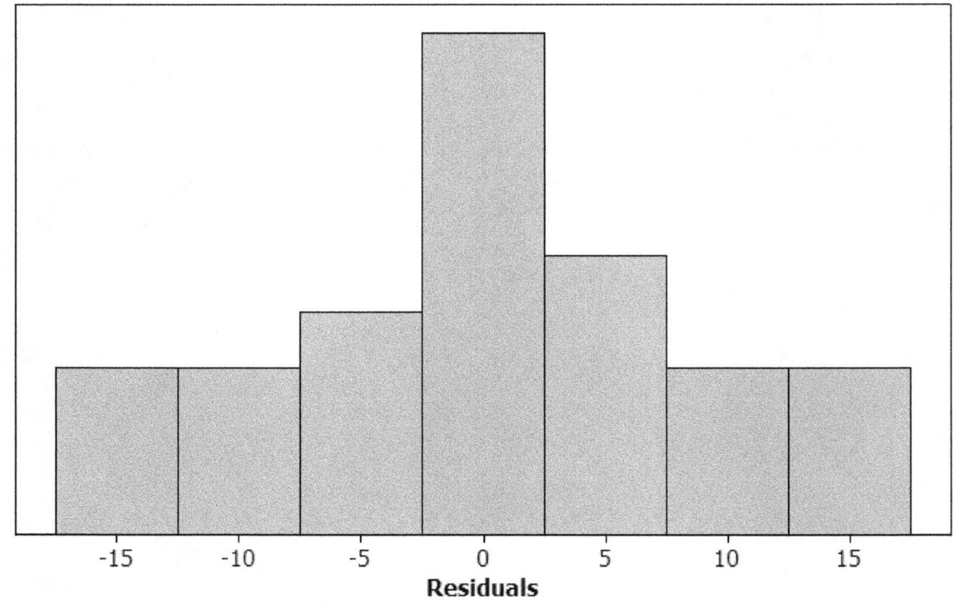

D. To check to see if our data provides evidence of equal standard deviations of y across all values of x, we plotted the residuals versus goals against. This plot is given in Figure 11.9.

FIGURE 11.9: Plot of residuals versus goals against for the 2011–2012 NHL season in Example 2.

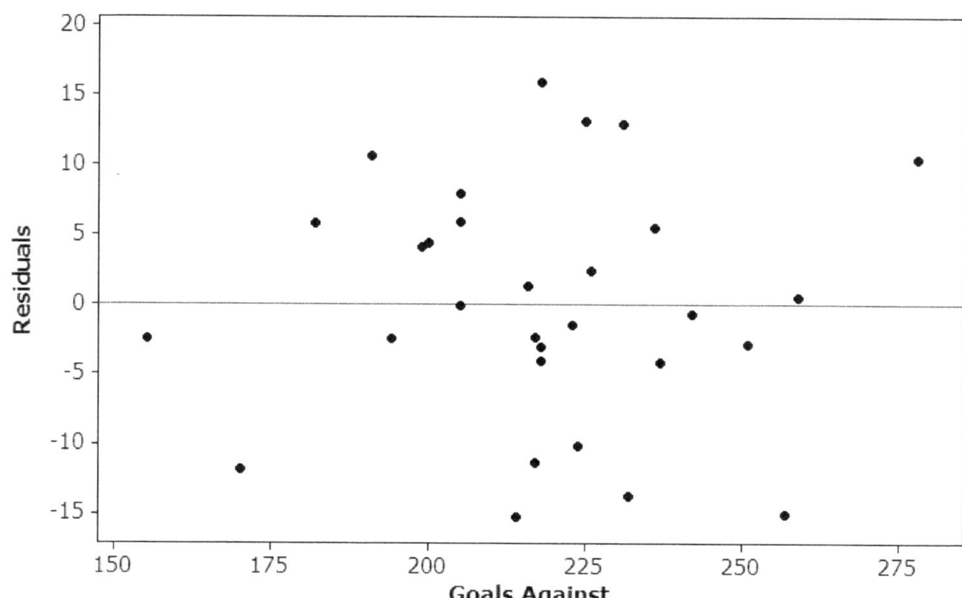

A plot of the residuals versus *goals against* shows a random pattern of residuals on both sides of 0. The points shows a nice, relatively even scatter about the zero-reference line. We can conclude the assumption of equal standard deviations for *points* across all values of *goals against* is met.

Step 3: We Set the Level of Significance at 0.05

Step 4: Calculate Test Statistic

We turn to Minitab and the annotated output in Table 11.4 to assist in the calculations.

```
The regression equation is Points = 159 - 0.308 Goals Against

                  y-intercept
Predictor            Coef    SE Coef       T       P
Constant           159.11      13.58   11.72   0.000
Goals Against    -0.30761    0.06182   -4.98   0.000
                    Slope      se(b)  T test   p-value for test
                                      statistic  of Ho: β = 0

R-sq = 46.9%  <-- coefficient of
                  determination R²
```

TABLE 11.4: Annotated Minitab regression output for NHL goals against and points data for the 2011–2012 season in Example 2.

From the output under the heading "Coef" are the regression line *y*-intercept and slope, respectively. The software rounds these values in providing the final regression equation of,

$$\text{Points} = 159 - 0.308 \text{ Goals Against}$$

Note that the direction of the slope and the correlation (r = - 0.685) given in our check of assumptions are both negative. This fulfills one of the correlation properties we discussed in Section 11.1 that the correlation and slope with share the same direction.

From the output, we can reconstruct the t-test statistic in Table 11.3 by,

$$t = \frac{b - 0}{se(b)} = \frac{-0.30761 - 0}{0.06182} = -4.98$$

With a sample size of n = 30, the degrees of freedom are n - 2 = 30 - 2 = 28.

Step 5: Calculate p-value

From the Minitab output in Table 11.4, we find a p-value of 0.000 for the test of Ho: β = 0. We can also find this using the T-Table, a portion of which is provided in Figure 11.10.

FIGURE 11.10: Portion of T-Table giving various *t* values with right-tail probabilities.

T-Table: t Distribution Confidence Interval and Critical Values

	Confidence Level					
	80%	90%	95%	98%	99%	99.8%
	Right Tail Probability					
df	$t_{0.10}$	$t_{0.05}$	$t_{0.025}$	$t_{0.01}$	$t_{0.005}$	$t_{0.001}$
1	3.078	6.314	12.706	31.821	63.657	318.289
2	1.886	2.920	4.303	6.965	9.925	22.328
.	....	....	....	....	....	....
.	....	....	....	....	....	....
28	1.313	1.701	2.048	2.467	2.763	3.408

Comparing the absolute value of the test statistic 4.98 to the t values in Figure 11.10 for df = 28, we see that 6.76 *exceeds* the largest t value of 3.408, corresponding to a 0.001 right-tail probability. Since our alternative hypothesis Ha is "not equal," we double this right-tail probability to conclude that the p-value would be less than 0.002 for our test of Ho: $\beta = 0$. This agrees with the Minitab output p-value of 0.000.

Step 6: Conclusion

With the p-value being less than 0.05, we reject the null hypothesis. We have enough statistical evidence to conclude that *goals against* is a significant linear predictor of *points* for the teams playing in the NHL. The output also includes the coefficient of determination R-sq, or R^2, of 46.9%. The interpretation of this R^2 is that 46.9% of the variation in points scored by teams in the NHL is explained by their goals against. (This also means that roughly half the variation in *points* is not explained by *goals against*. Possibly this remaining half—or at least a good portion of it—could be accounted for by including goals scored, the offensive piece.) Note that this R^2 of 46.9% coincides with one of the properties for R^2, that the measure be equal to the square of the correlation r when we have only the one predictor. Since we rejected the null hypothesis, we should include an interpretation of the slope, plus a confidence interval for the estimate. With the slope b of the regression equation of -0.308, we interpret this to mean that for every goal allowed by a team, their expected *mean* points will *decrease* by 0.308 points. Giving up a goal is equivalent to losing approximately a third of a point. A 95% confidence interval for the slope can be found by,

$$b \pm t_{0.025} SE(b)$$

where $t_{0.025}$ represents the t multiplier for a 95% confidence interval found in the T-table with degrees of freedom n - 2. From Table 11.4, using df of 28, we find the t multiplier to be 2.048 for a 95% confidence interval. Combining this with the Minitab output in Table 11.4, we have a 95% confidence interval for the slope,

$$b \pm t_{0.025} SE(b) = -0.30761 \pm 2.048(0.06182) = -0.30761 \pm 0.1266 = (-0.434, -0.181)$$

We would interpret this interval as,

We are 95% confident that the population slope falls from negative 0.434 to negative 0.181. Since the interval does not include zero, we would conclude that the population slope is not equal to zero at a 0.05 level of significance.

11.4: Confidence Intervals, Prediction Intervals, and the ANOVA Output

Interpreting the ANOVA Output for Regression

As we alluded to in Section 11.3, a regression analysis also includes a total sum of squares (SST) and sum of squared error (SSE). There is a third sum of squares, the **sum of squares regression** (SSR). These three measures are analogous to their counterparts we discussed for ANOVA in Chapter 9, with one main difference being that now our interest is in analyzing the *slope* of a line and not group-level means. In regression analysis, the SSR comes from the squared differences of the predicted y values and the mean of y. The equation is,

$$SSR = \sum (\hat{Y}_i - \bar{Y})^2$$

Since we know that SST = SSR + SSE and a *small* SSE compared to SST is indicative of x being a strong predictor of y, then in contrast, a *large* SSR compared to SST would also be indicative of a strong linear relationship. This gives us a second formula for finding R^2. We can use,

$$R^2 = 1 - \frac{SSE}{SST} \quad \text{or} \quad R^2 = \frac{SSR}{SST}$$

Table 11.5 includes the ANOVA output for the NHL data in Example 2.

Analysis of Variance

Source	DF	SS	MS	F	P
Regression	1	**1871.5**	1871.5	24.76	0.000
Residual Error	28	**2116.5**	75.6		
Total	29	**3988.0**			

TABLE 11.5: ANOVA output for NHL data in Example 2.

Using the summary measures under the Sum of Squares (SS) column, we get as R^2,

$$R^2 = 1 - \frac{SSE}{SST} = 1 - \frac{2116.5}{3988.0} = 0.469 \text{ or, } R^2 = \frac{SSR}{SST} = \frac{1871.5}{3988.0} = 0.469$$

Multiplying either by 100% results in the Minitab output R-sq of 46.9%.

Also in the output is the **mean squared error** (MSE). This is the value under the MS column in the Residual Error row. From Table 11.5, we find a value of 75.6, corresponding to the MSE for the NHL examples. The MSE is an estimate of the variance for y values for all subjects with the same value of x. The square root of the MSE is called the **residual standard deviation** denoted by S. The formula for MSE is,

$$MSE = \frac{SSE}{n-2} = \frac{\sum (Y_i - \hat{Y}_i)^2}{n-2}$$

and $S = \sqrt{MSE}$.

This standard deviation S differs from S_y, which is the standard deviation of all the y values in the set of data. The first one, S, is a measure of variability for all y values at equal values of x, while the latter, S_y, refers to the variability of the y values about the mean of y. As this relates to the NHL example, where $S = \sqrt{75.6} = 8.69$, this measures the variability in points for each level of goals against. The value of S_y represents the variability in points scored around the mean of points. For the NHL data, this standard deviation in points is 11.73. When we have a strong correlation, the variability in y across fixed values of x will be smaller than the overall variability in y. As was the case with the NHL with a moderate-to-strong correlation of -0.685, we see the residual standard deviation ($S = 8.69$) is quite a bit smaller than the standard deviation of y ($S_y = 11.73$).

Finding Intervals for Predicting y Outcomes

Inevitably, if we have a regression line that results in x being a significant linear predictor of y, we will want to use the line to predict an outcome for some x value. With the prediction, we would also want to provide, with some degree of confidence, a range for this outcome. However, in regression, we have two distinct intervals. One is a **prediction interval** that predicts where an individual y value will fall for some x value, while the second is a **confidence interval**, which estimates the population mean of y for some value of x. In terms of the NHL example, an example of a prediction interval would be where we would predict the points for a particular team allowing a specified number of goals, while a confidence interval would estimate the mean points for all teams allowing a specified number of goals. As with any interval, we need to incorporate some measure of error. The equations for finding these standard errors are quite complicated for an introductory statistics course and are better left for the software. We can, though, consider approximate 95% confidence and prediction intervals for some specified x value by,

Confidence interval: $\hat{y} \pm 2\left(\frac{S}{\sqrt{n}}\right)$

Prediction interval: $\hat{y} \pm 2S$

where $\hat{y}$ is the fitted (predicted) value of y for some specified value of x, S is the residual standard deviation found by taking the square root of MSE, and n is the sample size. In examining these two equations, we should notice that a prediction interval *will always be wider* than a confidence interval *except* when we have a perfect prediction; that is, we have no error. In such a rare situation, the confidence and prediction intervals are simply the predicted y values. Table 11.6 offers 95% confidence (CI) and prediction (PI) intervals for a team that allowed 200 goals against in a season, as calculated using Minitab.

```
Predicted Values for New Observations

New Obs     Fit   SE Fit          95% CI              95% PI
      1   97.59     1.94   (93.61, 101.57)   (79.34, 115.84)

New Obs   Goals Against
      1             200
```

TABLE 11.6: Minitab 95% confidence intervals (CI) and prediction intervals (PI) for a goals against of 200 for the NHL data in Example 2.

The value 97.59 below "Fit" is the predicted points for an observed goals against of 200. The SE fit value is the precise standard error Minitab is using to calculate the 95% confidence interval. What we have under "95% CI" is the 95% confidence interval for the expected mean points for teams allowing 200 goals in a season. The "95% PI" offers a 95% prediction interval for any one team giving up 200 goals in a season. Since the prediction was not perfect—there was error—the confidence interval is narrower than the prediction interval, as we expected. Using the approximate 95% intervals we calculate,

$$\text{Confidence interval: } \hat{y} \pm 2\left(\frac{S}{\sqrt{n}}\right) = 97.59 \pm 2\left(\frac{8.69}{\sqrt{30}}\right) = 97.59 \pm 3.17 = (94.42,\ 100.76)$$

$$\text{Prediction interval: } \hat{y} \pm 2S = 97.59 \pm 2(8.69) = 97.59 \pm 17.38 = (80.21,\ 115.18)$$

both of which are fair approximations to those produced in Minitab.

11.5: Some Cautions about Regression and Correlation

Although there are many factors that can influence regression results, two important ones to consider are **extrapolation** and sample size. Extrapolation is when we apply the regression line to an *x* value that is outside the range of *x* values used in finding the regression equation. In the NHL example, the data came from a full 82-game schedule for all 30 teams. The range of goals against was 155 to 278. Come next season, we want to apply the equation to estimate points for our favorite team after they have completed only half the games. If at this time they have only allowed 100 goals, the application of the regression equation would not be appropriate, as that *x* value of 100 is far outside the range of *x* values used to calculate the regression equation.

As to sample size, we discussed in Chapter 8 how a larger sample size can lead to rejecting the null hypothesis when the difference between the sample statistic and null value is small. When we apply this thought to the correlation, keep in mind the definition and properties of *r*. The value of *r* provides a measure of the strength of a linear relationship and when close to 0 indicates no linear relationship. When sample size increases, we raise the likelihood of finding a significant linear relationship, despite the correlation being quite small. That is, we could have a correlation *r* of 0.1 that produces a statistically significant result, whereby we conclude that *x* is a significant linear predictor of *y*. However, in the regression, this would result in an R^2 of only 1%, meaning that the predictor variable only explained one percent of the variation in the response variable! In practical terms, using that predictor is no better than using the mean of y.

Expressions and Formulas

1. Correlation r

$$r = \frac{1}{n-1}\sum\left(\frac{x_i-\bar{x}}{s_x}\right)\left(\frac{y_i-\bar{y}}{s_y}\right)$$

2. Regression line

$\hat{y} = a + bx$ where $\hat{y}$ is the predicted (or fitted) values of y, a is the y-intercept, and b is the slope

3. The least square estimates of the y-intercept and slope for the regression line

$b = r\left(\frac{s_y}{s_x}\right)$ and $a = \bar{y} - b\bar{x}$

4. Sum of squares in regression

$$\text{Sum of Squared Regression} = SSR = \sum(\hat{Y}_i - \bar{Y})^2$$

$$\text{Sum of Squared Error} = SSE = \sum(Y_i - \hat{Y}_i)^2$$

$$\text{Sum of Squared Total} = SST = \sum(Y_i - \bar{Y})^2$$

5. Coefficient of Determination R^2

$R^2 = 1 - \frac{SSE}{SST}$ or $R^2 = \frac{SSR}{SST}$

6. Population regression equation

$\mu_y = \alpha + \beta x$ where μ_y denotes the population mean of y; α is the population y-intercept estimated by a; and β is the population slope estimated by b.

7. Hypotheses for test of regression line slope

Ho: $\beta = 0$
Ha: $\beta \neq 0$

8. Test statistic for testing

$$t = \frac{b-0}{se(b)}$$

9. The mean square error, MSE and residual standard deviation, S

$MSE = \frac{SSE}{n-2} = \frac{\sum(Y_i - \hat{Y}_i)^2}{n-2}$ and $S = \sqrt{MSE}$.

10. approximately 95% confidence and prediction intervals for some specified x value

    Confidence interval: $\hat{y} \pm 2\left(\frac{S}{\sqrt{n}}\right)$

    Prediction interval: $\hat{y} \pm 2S$

Appendix

Z-Table: Standard Normal Cumulative Probabilities

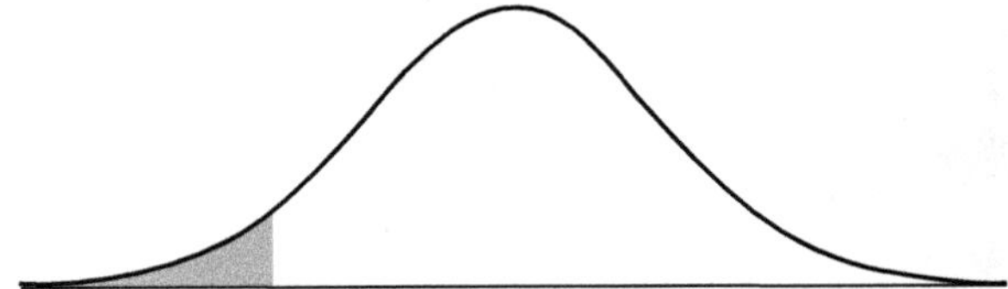

Cumulative probability (area to LEFT) of Negative Z-values

Z	0.00	0.01	0.02	0.03	0.04	0.05	0.06	0.07	0.08	0.09
-3.0	0.0013	0.0013	0.0013	0.0012	0.0012	0.0011	0.0011	0.0011	0.0010	0.0010
-2.9	0.0019	0.0018	0.0018	0.0017	0.0016	0.0016	0.0015	0.0015	0.0014	0.0014
-2.8	0.0026	0.0025	0.0024	0.0023	0.0023	0.0022	0.0021	0.0021	0.0020	0.0019
-2.7	0.0035	0.0034	0.0033	0.0032	0.0031	0.0030	0.0029	0.0028	0.0027	0.0026
-2.6	0.0047	0.0045	0.0044	0.0043	0.0041	0.0040	0.0039	0.0038	0.0037	0.0036
-2.5	0.0062	0.0060	0.0059	0.0057	0.0055	0.0054	0.0052	0.0051	0.0049	0.0048
-2.4	0.0082	0.0080	0.0078	0.0075	0.0073	0.0071	0.0069	0.0068	0.0066	0.0064
-2.3	0.0107	0.0104	0.0102	0.0099	0.0096	0.0094	0.0091	0.0089	0.0087	0.0084
-2.2	0.0139	0.0136	0.0132	0.0129	0.0125	0.0122	0.0119	0.0116	0.0113	0.0110
-2.1	0.0179	0.0174	0.0170	0.0166	0.0162	0.0158	0.0154	0.0150	0.0146	0.0143
-2.0	0.0228	0.0222	0.0217	0.0212	0.0207	0.0202	0.0197	0.0192	0.0188	0.0183
-1.9	0.0287	0.0281	0.0274	0.0268	0.0262	0.0256	0.0250	0.0244	0.0239	0.0233
-1.8	0.0359	0.0351	0.0344	0.0336	0.0329	0.0322	0.0314	0.0307	0.0301	0.0294
-1.7	0.0446	0.0436	0.0427	0.0418	0.0409	0.0401	0.0392	0.0384	0.0375	0.0367
-1.6	0.0548	0.0537	0.0526	0.0516	0.0505	0.0495	0.0485	0.0475	0.0465	0.0455
-1.5	0.0668	0.0655	0.0643	0.0630	0.0618	0.0606	0.0594	0.0582	0.0571	0.0559
-1.4	0.0808	0.0793	0.0778	0.0764	0.0749	0.0735	0.0721	0.0708	0.0694	0.0681
-1.3	0.0968	0.0951	0.0934	0.0918	0.0901	0.0885	0.0869	0.0853	0.0838	0.0823
-1.2	0.1151	0.1131	0.1112	0.1093	0.1075	0.1056	0.1038	0.1020	0.1003	0.0985
-1.1	0.1357	0.1335	0.1314	0.1292	0.1271	0.1251	0.1230	0.1210	0.1190	0.1170
-1.0	0.1587	0.1562	0.1539	0.1515	0.1492	0.1469	0.1446	0.1423	0.1401	0.1379
-0.9	0.1841	0.1814	0.1788	0.1762	0.1736	0.1711	0.1685	0.1660	0.1635	0.1611
-0.8	0.2119	0.2090	0.2061	0.2033	0.2005	0.1977	0.1949	0.1922	0.1894	0.1867
-0.7	0.2420	0.2389	0.2358	0.2327	0.2296	0.2266	0.2236	0.2206	0.2177	0.2148
-0.6	0.2743	0.2709	0.2676	0.2643	0.2611	0.2578	0.2546	0.2514	0.2483	0.2451
-0.5	0.3085	0.3050	0.3015	0.2981	0.2946	0.2912	0.2877	0.2843	0.2810	0.2776
-0.4	0.3446	0.3409	0.3372	0.3336	0.3300	0.3264	0.3228	0.3192	0.3156	0.3121
-0.3	0.3821	0.3783	0.3745	0.3707	0.3669	0.3632	0.3594	0.3557	0.3520	0.3483
-0.2	0.4207	0.4168	0.4129	0.4090	0.4052	0.4013	0.3974	0.3936	0.3897	0.3859
-0.1	0.4602	0.4562	0.4522	0.4483	0.4443	0.4404	0.4364	0.4325	0.4286	0.4247
0.0	0.5000	0.4960	0.4920	0.4880	0.4840	0.4801	0.4761	0.4721	0.4681	0.4641

Z-Table continued: Standard Normal Cumulative Probabilities

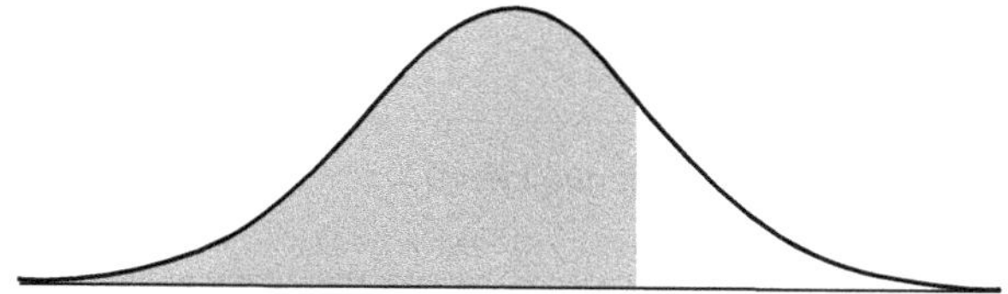

Cumulative probability (area to LEFT) of Positive Z-values

Z	0.00	0.01	0.02	0.03	0.04	0.05	0.06	0.07	0.08	0.09
0.0	0.5000	0.5040	0.5080	0.5120	0.5160	0.5199	0.5239	0.5279	0.5319	0.5359
0.1	0.5398	0.5438	0.5478	0.5517	0.5557	0.5596	0.5636	0.5675	0.5714	0.5753
0.2	0.5793	0.5832	0.5871	0.5910	0.5948	0.5987	0.6026	0.6064	0.6103	0.6141
0.3	0.6179	0.6217	0.6255	0.6293	0.6331	0.6368	0.6406	0.6443	0.6480	0.6517
0.4	0.6554	0.6591	0.6628	0.6664	0.6700	0.6736	0.6772	0.6808	0.6844	0.6879
0.5	0.6915	0.6950	0.6985	0.7019	0.7054	0.7088	0.7123	0.7157	0.7190	0.7224
0.6	0.7257	0.7291	0.7324	0.7357	0.7389	0.7422	0.7454	0.7486	0.7517	0.7549
0.7	0.7580	0.7611	0.7642	0.7673	0.7704	0.7734	0.7764	0.7794	0.7823	0.7852
0.8	0.7881	0.7910	0.7939	0.7967	0.7995	0.8023	0.8051	0.8078	0.8106	0.8133
0.9	0.8159	0.8186	0.8212	0.8238	0.8264	0.8289	0.8315	0.8340	0.8365	0.8389
1.0	0.8413	0.8438	0.8461	0.8485	0.8508	0.8531	0.8554	0.8577	0.8599	0.8621
1.1	0.8643	0.8665	0.8686	0.8708	0.8729	0.8749	0.8770	0.8790	0.8810	0.8830
1.2	0.8849	0.8869	0.8888	0.8907	0.8925	0.8944	0.8962	0.8980	0.8997	0.9015
1.3	0.9032	0.9049	0.9066	0.9082	0.9099	0.9115	0.9131	0.9147	0.9162	0.9177
1.4	0.9192	0.9207	0.9222	0.9236	0.9251	0.9265	0.9279	0.9292	0.9306	0.9319
1.5	0.9332	0.9345	0.9357	0.9370	0.9382	0.9394	0.9406	0.9418	0.9429	0.9441
1.6	0.9452	0.9463	0.9474	0.9484	0.9495	0.9505	0.9515	0.9525	0.9535	0.9545
1.7	0.9554	0.9564	0.9573	0.9582	0.9591	0.9599	0.9608	0.9616	0.9625	0.9633
1.8	0.9641	0.9649	0.9656	0.9664	0.9671	0.9678	0.9686	0.9693	0.9699	0.9706
1.9	0.9713	0.9719	0.9726	0.9732	0.9738	0.9744	0.9750	0.9756	0.9761	0.9767
2.0	0.9772	0.9778	0.9783	0.9788	0.9793	0.9798	0.9803	0.9808	0.9812	0.9817
2.1	0.9821	0.9826	0.9830	0.9834	0.9838	0.9842	0.9846	0.9850	0.9854	0.9857
2.2	0.9861	0.9864	0.9868	0.9871	0.9875	0.9878	0.9881	0.9884	0.9887	0.9890
2.3	0.9893	0.9896	0.9898	0.9901	0.9904	0.9906	0.9909	0.9911	0.9913	0.9916
2.4	0.9918	0.9920	0.9922	0.9925	0.9927	0.9929	0.9931	0.9932	0.9934	0.9936
2.5	0.9938	0.9940	0.9941	0.9943	0.9945	0.9946	0.9948	0.9949	0.9951	0.9952
2.6	0.9953	0.9955	0.9956	0.9957	0.9959	0.9960	0.9961	0.9962	0.9963	0.9964
2.7	0.9965	0.9966	0.9967	0.9968	0.9969	0.9970	0.9971	0.9972	0.9973	0.9974
2.8	0.9974	0.9975	0.9976	0.9977	0.9977	0.9978	0.9979	0.9979	0.9980	0.9981
2.9	0.9981	0.9982	0.9982	0.9983	0.9984	0.9984	0.9985	0.9985	0.9986	0.9986
3.0	0.9987	0.9987	0.9987	0.9988	0.9988	0.9989	0.9989	0.9989	0.9990	0.9990

T-Table: t Distribution Confidence Interval and Critical Values

	Confidence Level					
	80%	90%	95%	98%	99%	99.8%
	Right Tail Probability					
df	$t_{0.10}$	$t_{0.05}$	$t_{0.025}$	$t_{0.01}$	$t_{0.005}$	$t_{0.001}$
1	3.078	6.314	12.706	31.821	63.657	318.289
2	1.886	2.920	4.303	6.965	9.925	22.328
3	1.638	2.353	3.182	4.541	5.841	10.214
4	1.533	2.132	2.776	3.747	4.604	7.173
5	1.476	2.015	2.571	3.365	4.032	5.894
6	1.440	1.943	2.447	3.143	3.707	5.208
7	1.415	1.895	2.365	2.998	3.499	4.785
8	1.397	1.860	2.306	2.896	3.355	4.501
9	1.383	1.833	2.262	2.821	3.250	4.297
10	1.372	1.812	2.228	2.764	3.169	4.144
11	1.363	1.796	2.201	2.718	3.106	4.025
12	1.356	1.782	2.179	2.681	3.055	3.930
13	1.350	1.771	2.160	2.650	3.012	3.852
14	1.345	1.761	2.145	2.624	2.977	3.787
15	1.341	1.753	2.131	2.602	2.947	3.733
16	1.337	1.746	2.120	2.583	2.921	3.686
17	1.333	1.740	2.110	2.567	2.898	3.646
18	1.330	1.734	2.101	2.552	2.878	3.611
19	1.328	1.729	2.093	2.539	2.861	3.579
20	1.325	1.725	2.086	2.528	2.845	3.552
21	1.323	1.721	2.080	2.518	2.831	3.527
22	1.321	1.717	2.074	2.508	2.819	3.505
23	1.319	1.714	2.069	2.500	2.807	3.485
24	1.318	1.711	2.064	2.492	2.797	3.467
25	1.316	1.708	2.060	2.485	2.787	3.450
26	1.315	1.706	2.056	2.479	2.779	3.435
27	1.314	1.703	2.052	2.473	2.771	3.421
28	1.313	1.701	2.048	2.467	2.763	3.408
29	1.311	1.699	2.045	2.462	2.756	3.396
30	1.310	1.697	2.042	2.457	2.750	3.385
35	1.306	1.690	2.030	2.438	2.724	3.340
40	1.303	1.684	2.021	2.423	2.704	3.307
60	1.296	1.671	2.000	2.390	2.660	3.232
80	1.292	1.664	1.990	2.374	2.639	3.195
100	1.290	1.660	1.984	2.364	2.626	3.174
inf.	1.282	1.645	1.960	2.326	2.576	3.091

F-Table: F-values for Right Tail Probability = 0.05

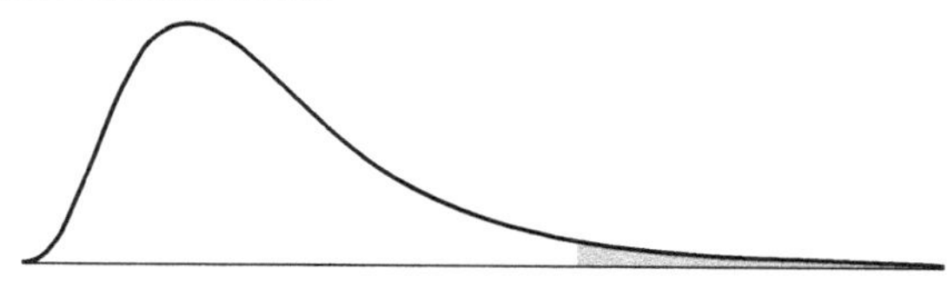

df_2 \ df_1	1	2	3	4	5	6	8	12	24	inf.
1	161.45	199.50	215.71	224.58	230.16	233.99	238.88	243.91	249.05	254.31
2	18.51	19.00	19.16	19.25	19.30	19.33	19.37	19.41	19.45	19.50
3	10.13	9.55	9.28	9.12	9.01	8.94	8.85	8.74	8.64	8.53
4	7.71	6.94	6.59	6.39	6.26	6.16	6.04	5.91	5.77	5.63
5	6.61	5.79	5.41	5.19	5.05	4.95	4.82	4.68	4.53	4.37
6	5.99	5.14	4.76	4.53	4.39	4.28	4.15	4.00	3.84	3.67
7	5.59	4.74	4.35	4.12	3.97	3.87	3.73	3.57	3.41	3.23
8	5.32	4.46	4.07	3.84	3.69	3.58	3.44	3.28	3.12	2.93
9	5.12	4.26	3.86	3.63	3.48	3.37	3.23	3.07	2.90	2.71
10	4.96	4.10	3.71	3.48	3.33	3.22	3.07	2.91	2.74	2.54
11	4.84	3.98	3.59	3.36	3.20	3.09	2.95	2.79	2.61	2.40
12	4.75	3.89	3.49	3.26	3.11	3.00	2.85	2.69	2.51	2.30
13	4.67	3.81	3.41	3.18	3.03	2.92	2.77	2.60	2.42	2.21
14	4.60	3.74	3.34	3.11	2.96	2.85	2.70	2.53	2.35	2.13
15	4.54	3.68	3.29	3.06	2.90	2.79	2.64	2.48	2.29	2.07
16	4.49	3.63	3.24	3.01	2.85	2.74	2.59	2.42	2.24	2.01
17	4.45	3.59	3.20	2.96	2.81	2.70	2.55	2.38	2.19	1.96
18	4.41	3.55	3.16	2.93	2.77	2.66	2.51	2.34	2.15	1.92
19	4.38	3.52	3.13	2.90	2.74	2.63	2.48	2.31	2.11	1.88
20	4.35	3.49	3.10	2.87	2.71	2.60	2.45	2.28	2.08	1.84
21	4.32	3.47	3.07	2.84	2.68	2.57	2.42	2.25	2.05	1.81
22	4.30	3.44	3.05	2.82	2.66	2.55	2.40	2.23	2.03	1.78
23	4.28	3.42	3.03	2.80	2.64	2.53	2.37	2.20	2.01	1.76
24	4.26	3.40	3.01	2.78	2.62	2.51	2.36	2.18	1.98	1.73
25	4.24	3.39	2.99	2.76	2.60	2.49	2.34	2.16	1.96	1.71
26	4.23	3.37	2.98	2.74	2.59	2.47	2.32	2.15	1.95	1.69
27	4.21	3.35	2.96	2.73	2.57	2.46	2.31	2.13	1.93	1.67
28	4.20	3.34	2.95	2.71	2.56	2.45	2.29	2.12	1.91	1.65
29	4.18	3.33	2.93	2.70	2.55	2.43	2.28	2.10	1.90	1.64
30	4.17	3.32	2.92	2.69	2.53	2.42	2.27	2.09	1.89	1.62
40	4.08	3.23	2.84	2.61	2.45	2.34	2.18	2.00	1.79	1.51
60	4.00	3.15	2.76	2.53	2.37	2.25	2.10	1.92	1.70	1.39
100	3.94	3.09	2.70	2.46	2.31	2.19	2.03	1.85	1.63	1.28
inf.	3.84	3.00	2.60	2.37	2.21	2.10	1.94	1.75	1.52	1.00

Chi-Square Table: Chi-Square (X^2) Values for Various Right Tail Probabilities

Right Tail Probability

df	0.250	0.100	0.050	0.025	0.010	0.005	0.001
1	1.32	2.71	3.84	5.02	6.63	7.88	10.83
2	2.77	4.61	5.99	7.38	9.21	10.60	13.82
3	4.11	6.25	7.81	9.35	11.34	12.84	16.27
4	5.39	7.78	9.49	11.14	13.28	14.86	18.47
5	6.63	9.24	11.07	12.83	15.09	16.75	20.52
6	7.84	10.64	12.59	14.45	16.81	18.55	22.26
7	9.04	12.02	14.07	16.01	18.48	20.28	24.32
8	10.22	13.36	15.51	17.53	20.09	21.96	26.12
9	11.39	14.68	16.92	19.02	21.67	23.59	27.88
10	12.55	15.99	18.31	20.48	23.21	25.19	29.59
11	13.70	17.28	19.68	21.92	24.72	26.76	31.26
12	14.85	18.55	21.03	23.34	26.22	28.30	32.91
13	15.98	19.81	22.36	24.74	27.69	29.82	34.53
14	17.12	21.06	23.68	26.12	29.14	31.32	36.12
15	18.25	22.31	25.00	27.49	30.58	32.80	37.70
16	19.37	23.54	26.30	28.85	32.00	34.27	39.25
17	20.49	24.77	27.59	30.19	33.41	35.72	40.79
18	21.60	25.99	28.87	31.53	34.81	37.16	42.31
19	22.72	27.20	30.14	32.85	36.19	38.58	43.82
20	23.83	28.41	31.41	34.17	37.57	40.00	45.31
21	24.93	29.62	32.67	35.48	38.93	41.40	46.80
22	26.04	30.81	33.92	36.78	40.29	42.80	48.27
23	27.14	32.01	35.17	38.08	41.64	44.18	49.73
24	28.24	33.20	36.42	39.36	42.98	45.56	51.18
25	29.34	34.38	37.65	40.65	44.31	46.93	52.62
26	30.43	35.56	38.89	41.92	45.64	48.29	54.05
27	31.53	36.74	40.11	43.19	46.96	49.65	55.48
28	32.62	37.92	41.34	44.46	48.28	50.99	56.89
29	33.71	39.09	42.56	45.72	49.59	52.34	58.30
30	34.80	40.26	43.77	46.98	50.89	53.67	59.70
40	45.62	51.80	55.76	59.34	63.69	66.77	73.40
50	56.33	63.17	67.50	71.42	76.15	79.49	86.66
60	66.98	74.40	79.08	83.30	88.38	91.95	99.61
70	77.58	85.53	90.53	95.02	100.4	104.2	112.3
80	88.13	96.58	101.8	106.6	112.3	116.3	124.8
100	109.1	118.5	124.3	129.6	135.8	140.2	149.5

Index

SYMBOLS

A

B

C

D

E

F

G

H

I

L

M

N

O

P

Q

R

S

T

U

V

Z

www.ingramcontent.com/pod-product-compliance
Lightning Source LLC
Chambersburg PA
CBHW061411210326
41598CB00035B/6174

* 9 7 8 1 6 2 1 3 1 6 4 0 4 *